Doors

Liam Grey

Dedication

To Hassan, who inspired me. I gave you the ending that you deserve, and I will never forget you.

To every heart living in fear...

To everyone who refused to be in the shadows...

To every soul that has awakened from a long slumber...

To everyone who thinks they are incapable of living happily...

To every person who wants to be the self that dreams about...

To every person with scars, be proud. Your scars are your medals...

To every person who fought or is still fighting in this life...

You create your own destiny.

Acknowledgment

This fantastic endeavor would not have been possible without the utmost support of my dear Yousuf, Jaime Lee Rolland, Aldo Jaku and Sam Phan, Bernadeth Bangod and Michael Failla. It has been like a dream, and I would like to acknowledge the efforts of all those who stood in my support in making this dream of mine come true.

CONTENTS

About the Author

In 1984, in Baghdad, Iraq, Liam Grey (Middle Eastern mix) was born amid the regional rifts to pursue his ideal life. Unlike other authors, Liam has embraced challenges with open arms since adolescence as the prevalent norms in his region did not suit his ideas. He is a firm believer in living and letting live. He preaches that life should be enjoyed no matter what the circumstances are.

Growing up, drama and the supernatural have been his forte; his life revolves around these two things. However, as Liam is a creative genius, he always plays with multiple thoughts to develop the most extraordinary ideas, making him stand out in the crowd; his friends would surely nod to that.

Before permanently settling in Toronto, Liam hustled to live between Iraq and Turkey for some time. So, it wouldn't be wrong to say that his personality reflects a blend of the middle east and the west.

He never settles for less and pursues his unique interests with sheer dedication. For instance, he has graduated with a degree in Agriculture Engineering and a Masters in Soil Physics, and is now an author too.

Not just his personal life, his professional life is diverse too. The guy wants to try everything. He served as a teacher for seven years in a university and then moved to ORAM (Organization for Refuge, Asylum, and Migration).

Talking about his witty character, he possesses hilariously dry and dark humor. He would nail (a slang for troll – so that you know) anybody with his words, leaving them shook and sharing laughter with everyone. What's amazing is that this humor regurgitates mysteriously because if there is an attempt to troll Liam, uh-huh, then brace yourself for a plethora of kill shots coming your way in the form of hilarious jokes.

To let it all out, though, he often writes poems in Arabic that serve as the advocates of his feelings, the ones he prefers to keep to himself.

Preface

Based on the life of an ambitious man, Hassan, who was born in Baghdad and moved to Canada amid the regional rifts to pursue his ideal life. But, as he was born in a conservative state, being a gay Muslim made him undergo testing times.

He possesses a charismatic personality; he is a handsome man with green eyes blessed with the gift of love and friendship. Since he was gay, and on top of that, a Muslim, he could not afford to live in Baghdad anymore. Moving to Canada was not it, though. He found love there, made great friends, and brought positive energy and spark wherever he went. As he progressed in life, hardships continued to follow him. He got diagnosed with HIV, which he got from his lover, but that did not slow him down since he was resilient and ready to take on anything that crossed his path.

Due to an influencing attribute, everybody looked up to him. He moved forward in his life, made others do the same, and was never disappointed with his past. It was when he found love again, something he had not anticipated after contracting HIV, and went on living the rest of his life with him, but at the cost of being kidnapped and getting tortured for days and eventually losing a close friend.

Because of his dedication and devotion toward his loved ones and the irrepressible persona, Hassan continued to live the rest of his life overcoming the hurdles not just for himself but also for others to bring comfort and happiness in their lives and promote compassion.

Chapter 1: Family

From Baghdad to Canada

"The greatest healing therapy is friendship and love." - Hubert Humphrey

Hassan was fortunate to have both in his life, love and friendship. However, it had been an extremely tough journey for him. A gay Muslim from Iraq, Hassan was bound to go through anguish. Yet, it was not how one would expect the suffering to unfold.

A handsome man with eyes that represented a reflection of the greenest of seas, Hassan was a warm and friendly person. Everyone he interacted with would gravitate towards his warmth and welcoming smile. He had long brown hair that fell charmingly on his shoulders. The long hair was a way for Hassan to compensate for his height since he was only 168 centimeters tall.

He grew up in Baghdad and had a fairly happy childhood. He came from a wealthy family. It was his home but now, a mere thought of Baghdad would only bring back extremely vivid memories for Hassan. It was the place where he not only lost the love of his life but his entire family. His relatives who returned his father in a box because his father transferred all his belongings under Hassan's name. This enraged Hassan's grandfather who then killed his own son. They later kidnapped Hassan with his boyfriend. His boyfriend was kidnapped, raped, tortured and ruthlessly killed by his relatives. Somehow Hassan managed to escape but when he returned, he found his family under the wreckage.

Hassan was shattered and could not bear staying in the city which took everything from him. Additionally, he knew that his sexual preferences would not be tolerated in Baghdad. Thus, he decided to move to Turkey. All these fears and tragedies impacted Hassan's mental health severely. He faced extreme anxiety in his usual routine which he had never experienced before. Furthermore, it became a huge task for him to even conversate with people as his mind was stuck in dark clouds of thoughts.

Fighting with his depression and anxiety of being in a new country, Hassan was having a hard time adjusting. Hassan had filed for asylum and moved to Manisa, a city in Turkey, as a refugee. He was not able to capture the splendor of Turkey due to the revulsions of his past.

Nonetheless, Hassan was brave and knew he had to fight the demons within him to move on in life.

Hassan was academically gifted which he had proven by becoming a chemical engineer. He received an offer from a reputable company in Canada and decided to move there. He wanted to distance himself as much as he could from Iraq, and Canada, being at the other end of the globe, gave him the perfect opportunity to do so.

On his flight to Canada, Hassan met a guy named Ethan. Like all others, Ethan gravitated towards him to the extent that you could call it love at first sight. Coincidently, Ethan was working for the same company that offered Hassan the job. They had a pleasant conversation which quickly turned into a friendship.

Since Hassan was new to the country, Ethan offered him a ride to his new place. John had arrived to pick Ethan from the airport. John and Ethan had a passionate and troublesome past. However, since Ethan was falling for Hassan, he did not feel like introducing him to John at this point in time; instead, he ignored John and acted as if he didn't see him at the airport. John felt humiliated and embarrassed at this behavior. He noticed the way Ethan was looking at Hassan and the way he was smiling while talking to him. John still had feelings for Ethan and so, he got jealous and decided to steal Hassan away from him, thinking of it as a way to avenge the disappointing end to their past relationship. Little did John know; the revenge would soon turn into a love story between Hassan and himself.

***

Hassan had settled well in Toronto; he had a few friends and was recovering quite well from his past. Two years had passed since he had met Cara. Cara was in the same boat as Hassan. She had recently lost her husband. The trauma of losing a loved one deeply affected her mental and physical health. She was also pregnant at the time and due to the anguish, it ended in a miscarriage.

It was a Sunday afternoon when Hassan would go to this café to enjoy his favorite dish, Greek salad. The café had a middle eastern ambiance to it which reminded him of his college days back home, hanging out with friends indulging in coffees and snacks. One day, he was eating his salad, when he noticed a woman sitting by herself staring at the television screen, while slowly sipping on a cup of coffee. She seemed numb and being the kind of person that Hassan was, he rushed over to talk to her.

"I couldn't help notice the sadness on such a pretty face," Hassan tried comforting Cara.

"I'm not someone to hit on, especially not at this time!"

"I would hit on you if you were a guy but since you're not maybe we can be friends?"

Cara embarrassingly replied, "I'm so sorry for judging you, I'm Cara, nice to meet you!"

After a heart to heart with Cara, Hassan had won her over as a good friend. She loved how he was able to see her despondency from afar. His warm and gentle nature welcomed Cara into his life. Additionally, Hassan had also shared his past to comfort her which made the two connect and help get close.

"I thought I was going through a lot but your story has shaken me to the core," said Cara.

"God only lets you suffer as much as you can handle, some people can handle more than others, however, the key is to allow yourself to suffer and move on before it starts to consume you," said Hassan comfortingly.

***

A beacon of light

"There is only one happiness in this life, to love and be loved." - George Sand

Love is exactly what Hassan had found. Rising from jealousy turning into love, John was forced to fall for Hassan's nature. He had never met a man who was as caring and sensitive as Hassan. He always made sure that John was truly happy. From picking out his clothes, cooking him meals and being there whenever John needed him, Hassan did it all. John found Hassan as the one man he could truly depend on. Thus, John confessed his love to Hassan.

***

Five years had passed since Hassan landed in Canada. His life was filled with elation as he found love and friendship again. John, Cara and Hassan were all living together now. John was now in a relationship with him while Cara had become his best friend.

The happiness he derived from the two helped him fight his past, and same was the case for Cara. That is the power of love, if it can put you at your lowest it most definitely can bring you back to your highest as well.

***

It was a sublime Friday night as the clouds had taken control of the skies while the moon managed to peep through the clouds to cast a beacon of light that seemed like a ray of hope for Cara and Hassan. All three of them were in the kitchen area, Cara continued cooking while the other two sat there on the counter stools in despair.

"Chop chop Cara, I am dying!" cried John as his stomach had starting eating itself.

"Hold your horses!" exclaimed Cara as she was too focused on her art.

Hassan intervened by grabbing John's hand, "Let our love fill the empty void in your stomach." All three giggled but seeing the two hold hands made Cara feel a little out of place. She told them how she wanted a relationship like the one they have. Hassan quickly jumped to make her feel better, "One day you will have it all. You are one of the nicest souls I have ever met and you deserve the very best! The perfect guy is out there dying to catch a glimpse of you. Let time work its magic!"

Hassan always knew the right words to say at the right time. He was gifted with high emotional intelligence which allowed him to easily gauge others' feelings and make them feel better. That is the reason why everybody naturally gravitated towards him. He would make people feel welcomed through his generous smiles and strong listening skills. He would give them all his attention making them feel like the most important person in his life.

Cara informed Hassan and John that her brother, Brian, would be coming over given that they were okay with it. "You live here just as much as we do, you can have visitors whenever you want," comforted Hassan. Cara went on to explain the hesitation, "Brian is a little different... He is quite challenging to deal with!" Hassan replied with confidence, "I have dealt with a lot and with me on John's side, we can deal with a lot more!" All of them started to laugh. Cara was right though; Brian had gone through a lot and was mentally unstable.

Growing up, Brian faced a difficult life as he was bullied most of his childhood. The bullying had devastating impacts on his personality. He became awfully hungry for any kind of revenge just so he could get back to anyone who he thought had wrong intentions for him. He also had a rough time with a therapist, Manauris, who would expose his clients to their fears and torture them until they got over their fears. Brian was, thus, a little troubled …

Dinner was ready and they all gathered around for a prayer. Cara offered the prayer and the room echoed with 'Amen'. They were about to lift their forks and knives to take a bite of the mouth-watering steaks when Cara said, "I love Fridays! We all get a chance to enjoy a meal together like a family." John raised an eyebrow and said, "I thought we were a family." All of them smiled as they dug into their meals.

After dinner, Hassan proceeded to washing dishes while John gazed at his beauty. Cara was watching from afar, enjoying a cup of tea. John approached from behind and wrapped his hands around Hassan's waist. "I love you, shorty," John whispered into his ears, followed by kisses on the neck. Hassan blushed as he replied, "I love you more!" Cara decided to leave the love birds alone and so vanished to her room. Hassan finished up the dishes and took hold of John, "Five years I have been with you, yet everyday my love only grows for you. I cannot imagine a life without you!"

***

Cara, while watching her favorite show 'Mom', could not help think about how she wanted a relationship like the one her roommates have. Usually, she watched the show with Hassan but she didn't want to ruin the night for John and him. Luckily, Anna Faris was able to distract her with some good laughs. Suddenly, a knock at the door caught her by surprise.

"You're watching 'Mom' without me?" asked Hassan in shock.

Cara replied, "I am sorry, I thought you're busy with John, you know..."

Hassan, with a smile on his face, replied, "John knows I cannot miss watching 'Mom' with you."

John also ended up in the same room and all three enjoyed the show for a couple hours until it was time for bed. Hassan and John headed to their room whilst singing their love for each other. As soon as they entered the room, they started kissing passionately and John pushed Hassan onto the bed. John got on top of Hassan, wrapping his legs around him and started kissing him again.

"I can stay like this forever," proclaimed John. "Kissing you is the best feeling in the world, I love your soft lips so much, they feel like butter melting in my mouth!"

Hassan replied, "You cannot fathom the love I have for you. No words can ever describe what I feel about you. The only words that can do a little bit of justice would be that I love you more than I love myself!"

They continued making love while talking about their future together and spending their lives full of happiness. They fell asleep making love and talking about their beautiful future.

The next morning, John woke up cheerful as a little boy on his birthday and decided to make breakfast for Hassan and Cara. John proceeded to wake Hassan up, asking him to join him in the shower. After the shower, the three of them got together for breakfast.

Cara noticed John's enthusiasm and said, "John, you seem super active today. May I ask why? Don't tell me you guys did it this morning?"

John, with a smirk on his face, replied, "Cara, you are making Hassan blush, and yes, we did it last night and this morning in the shower. Sorry, Cara, but we are in love!"

Cara started making fun of Hassan for blushing even though they have been together for so long. John quickly came to Hassan's rescue, "It is because you are an innocent person that life did not ruin, your heart is pure. That is why I love you. I have never met anyone like you. I am happy you are the man I love and adore." After breakfast, Cara and John started cleaning the house while Hassan went to the basement. They both talked about how much they loved him and how their life was miserable before him. Cara joked that if Hassan was not in love with John, she would definitely steal him away from him. Cara then asked John to buy her some things from the supermarket as she was expecting her brother and a few friends for a dinner. John agreed and left.

A twist in the tale

"A deep man believes that the evil eye can whither, the heart's blessing can heal, and that love can overcome all odds." - Ralph Waldo Emerson

Cara's brother, Brian, had finally made it to Toronto. Brian had long bangs of dark black hair flowing down his forehead. His nails were covered with black nail polish and eyes filled with Mascara. His appearance was significantly influenced by Gothic style and his attitude could be perceived as that of Michael Myers'; silent, daunting and eerie.

On reaching Cara and Hassan's place, he met his sister and instantly felt that something was bothering her. Cara told him she was just tired from all the cleaning. Brian seemed to have something up his sleeve, it seemed like he wanted to cause distress right away. "Obviously you're troubled. The reason is you're living with a Muslim from the middle east! They're barbaric, uneducated and ill-mannered!" Cara was extremely aggravated at Brian, "You're so racist! Hassan is one of the nicest person I have ever met! If you don't like him, please leave now!" Unfortunately, Hassan heard everything. Brian promised to behave and told Cara he was only there for her.

Hassan was extremely upset but did not want to talk about it. He decided to keep it to himself. However, John and Cara both noticed that something was troubling him. Cara asked John to take care of him, so he went to their room and decided to give him the present that he bought earlier to make him feel better. "You know I don't wear anklets," Hassan gently refused. John asked him to wear it since it was for men and he really wanted to see him wearing it. He had managed to distract Hassan for a while.

***

The guests started to come. Brian got dressed and went into the living room., and saw man he could not take his eyes off of. He introduced himself to which the man replied, "I am the barbaric and uneducated middle eastern!" Hassan walked away leaving Brian with so many questions. Brian quickly investigated if Cara had told Hassan. They discussed the matter and it was evident that he overheard them.

"Can you please clear the air between us, I really like him…"

Cara replied in anger, "Stop it! You have already messed with our friendship; he has a boyfriend! Do not mess with that! Promise me you will stay away from him?"

Brian promised to keep his distance. Meanwhile, Cara went over to Hassan to apologize. Hassan knew it was not her fault and so comforted her. Brian managed to creep up from behind and put his hand on Hassan's back. Hassan promptly asked Brian to move his hand but John had already seen it. John went over and kissed Hassan to mark his territory.

During the dinner, Brian could not stop gawking at Hassan and John. He felt extremely envious as he never found a relationship like theirs. Brian had an evil eye for them and he sure was planning something. Hassan was always at the receiving end for jealousy-based revenge. First with John and now Brian. However, this time, the events would take a different turn…

***

Everybody left after dinner except for Brian. The four sat down for a round of drinks. Brian felt out of place with them as he knew he was the 'outsider'. He jumped into the conversation by asking them if they knew about any good rental apartments. They were not aware of any such apartments and so Brian asked to stay with them for a while. He had the audacity to offer rent for the room but it was clear that Hassan was not interested in keeping him for long. John could feel Hassan's reservation for the room and decided to stay quiet.

After everyone dispersed, John asked Hassan why he wouldn't rent out the room. Hassan explained that he wanted that room empty for John's mother so that she could come by and stay whenever she wanted to. Hassan mentioned he was meeting a friend tomorrow and excused himself. However, he could not sleep due to the encounter with Brian.

The next morning, Hassan was not feeling good which was noticed by Cara. Hassan told her that he felt something bad will happen and so he wasn't been able to rest properly. Cara asked Hassan to skip work and offered to cover for him. After breakfast, John and Cara left for work.

Brian and Hassan happened to bump into each other in the kitchen. Brian noticed Hassan drinking a glass of orange juice made by Cara, and tried to blame him for using her but Hassan brushed him away by telling him not to interfere in their internal matters as he is not aware of their relationship. Furthermore, Hassan told him, "Do not ever think about doing it again, thank God John was asleep. You need to keep yourself in check!" Brian got offended at this statement and started arguing to which Hassan replied, "I know who you are, does the name Manauris bring back any memories?"

Brian was shocked at Hassan's knowledge as he thought nobody would know about his past. Extremely upset, he decided to take some action against Hassan. He started plotting about using John to ruin their relationship. Meanwhile, Hassan went to meet an old friend and could not stop thinking about last night and when he should tell John about Brian. His friend warned him…

Hassan explained, "You are so wrong about him! He's the perfect match for me and I will marry him, I love him so much!"

The friend replied, "He probably loves you too but trust me he is not right for you."

Hassan told him about being in a dark place after the friend left and that John was the one who rescued him. "Getting over you was testing for me, I can never forget what happened to us," said Hassan. Hassan's friend departed leaving him with a bunch of questions.

***

John called Hassan and informed him that they will be visiting his mother for dinner for which Hassan was excited. Hassan went home and decided to take a nap.

He sees himself on the roof of a building and John coming from behind. John pushes him off the edge. Hassan starts falling and sees water at the bottom. As he is about to hit the water, he closes his eyes. "Open your eyes," someone whispers. Hassan finds himself chained up in a dirty old room. He sees someone in a black hoodie approaching him. He sees himself, it's his biggest fear, his own loneliness. The hooded person tells Hassan, "He will break your heart like before! This time you will leave him for good.

Hassan finally woke up to John kissing him on the cheeks, "I am home shorty!" The two go for a shower and get dressed in their finest clothes. They head over to John's mother's house while Cara started preparing dinner for Brian. Taking the advantage of being alone with his sister, Brian started belittling her roommates and asked Cara to move in with him when he finds a place. He told her to get a life as she works and lives with the same person.

Brian goes on, "How will you find a boyfriend this way?"

To which Cara hysterically said, "You should not talk about boyfriends! You ruined my life three years ago, please, don't do it again."

Brian then promised and ensured that he only wants the best for his sister now.

***

That night John had a nightmare too. He saw Hassan kissing a stranger through a mirror and screamed. Hassan looked at him and laughed. John got frustrated and tossed a chair at the mirror. After the mirror shattered, John saw Hassan locked up in chains, bleeding intensely with dirt all over him. He also sees a knife in his hand and notices himself walking towards Hassan. He tries to stop himself but is unable to. He slits Hassan's throat and wakes up due to the gruesome visual. Hassan was already up by then and comforted John back to sleep.

Barbeque for dinner and a date on the side, please!

"Barbeque may not be the road to world peace, but it's a start." — Anthony Bourdain

The next day, Cara, Brian and Hassan decided to have a barbeque at their place and invited their friends over. Brian insisted Cara on bringing a date over which he found on Grindr. Cara rejected the idea of having a stranger at their house.

She wasn't feeling well that day and so, called Hassan to get a day off. After the call, she headed to her room and took a picture out from her drawer. As she examined the picture, it triggered memories from her past. Massive tears started flowing from her eyes as she put the picture close to her heart. Hassan's call interrupted the moment.

Hassan figured something was bothering Cara and asked, "You sound low, are you crying?"

Cara struggled to say, "Yes I am, how did you know?"

Hassan replied, "I know you very well. I cannot watch you fall again. I will take an early off and we will go shopping together!'

After the call, she fell back in her world of memories. All she could think about was how her dreams had shattered. She was alone with memories that daunted her. She would feel the sadness and pain at every moment in her life. She would spend hours just staring at the ceiling and crying. Luckily, Hassan came to interrupt it again by taking her shopping. On their trip, Hassan explained, "When we feel lonely, we tend to beat ourselves up. We start feeling like we don't belong or feel rejected by others. An isolated space is the perfect breeding ground for harmful thoughts. Remember that I am always here for you. You need to let the past go and move on in life." Cara nodded in approval but all she could think about was how words cannot erase the pain. However, she kept thinking how Hassan was always there for her and decided to try to move on for his efforts. She put a smile on her face and tried to face the world in a more positive light.

Hassan ended up buying her a dress to make her feel better and some other things for the barbeque party. Throughout the trip, Hassan tried everything to make Cara laugh. Laughing was the last thing Cara expected but Hassan, as always, managed to pull it out from her. On their way back, Hassan spotted an old friend, Kyle. He met Kyle and noticed how he was eyeing Cara and so, decided to invite him over. Cara asked Hassan why he did so as she could sense him trying to set her up. Hassan explained that he wanted her to move on in life and that Kyle would be the right guy for that as he has also suffered in his past relationship.

***

In the evening, Cara and Hassan were preparing for the barbeque when Kyle rang the doorbell. After seeing Cara, Kyle wasn't able to take his eyes off her. Hassan warned him not to mess around as she was family. Kyle mentioned that this is the first time he had felt something after his breakup.

Hassan helped Kyle secure a date with Cara over the weekend. They continued preparing for the barbeque while Kyle kept making moves on Cara. John entered the kitchen and noticed Cara in a different light. Hassan quickly took John outside to prepare the grill, hoping to not ruin the moments for Cara and Kyle. John knew what he was doing and said, "I love you, you pimp," making Hassan laugh.

Meanwhile, Cara and Kyle were getting closer by the moment. He would try to hold her hand, make her laugh and was in the middle of a meaningful conversation, when John and Hassan joined them as well. Hassan was ecstatic that his plan had worked. John whispered to him, "The pimp must be happy," followed by a kiss. Cara looked at the two wondering when she would have that. To her surprise, Kyle swooped in and gave her a kiss as well.

John's mother, Marguerite, and Ethan also arrived. Ethan was Brian's Grindr date and upon finding this out, Hassan warned Ethan to stay away from Brian. Ethan wanted to know why but Hassan told him that they will discuss it later.

Love prevails

"Love never claims, it ever gives; love never suffers, never resents, never revenges itself. Where there is love there is life; hatred leads to destruction." - Mahatma Gandhi

Hassan was delighted at the sight of his loved ones around him. He could not have asked for anything more. He looked over at John with utmost endearment. "My heart only beats for you," whispered Hassan into John's ear, slightly nibbling on his ear. "Well, prove it in bed tonight," said John with sheer excitement. The guests had started to leave and the night was wrapped up as one of the best nights in a long time, for all of them. John picked Hassan in his arms and took him upstairs, gently laying him down on the bed. He got on top and went towards his neck. The night was one to remember for the two of them…

***

Hassan and John's connection had become very strong by now, their love seemed to have no boundaries. That is the beauty of unconditional love. It is the choice to love no matter what comes your way. You love the other at their best and worst. You must accept all the flaws and insecurities while loving them wholeheartedly at the same time. However, it is extremely rare to find unconditional love, no matter how much one loves, it usually comes with its conditions. John and Hassan were fully content with what they had, not knowing what the future holds for them……

Chapter 2: Moving on and letting go

Another day, another problem

"What is life, after all, but a challenge?" – Warren Spahn

A month and a half later, after the barbeque, Brian had finally found a place for himself. It was the kind of place that matched Brian's personality. A small dusky room with not much furniture around. Just a mattress and a side table, a hollow room filled with darkness. The good thing was that, now, Brian would not be able to spread as much hate as he tried to do all the time whilst living with Hassan and friends. So, things at the prior household had started to normalize.

Hassan, on the other hand, had to deal with another co-worker, Daniel. Daniel was a young man in his early thirties. He was fairly white with blonde hair and looked like he was always in a hurry. He would start twitching when he was not moving around. It seemed like he was extremely anxious as he would always be doing something, he could never sit somewhere peacefully.

It was a bright sunny day and Hassan was feeling all the vibes to have a productive day at the office. He was feeling good and confident. His love life was progressing significantly. His relationship with Cara had become even better than before, given that Cara was now much more content in her life having Kyle around. The best thing was that Brian had moved out so there was not much room for negativity left in Hassan's life.

Hassan was sorting through files at work when Ethan suddenly barged into his office. Ethan quickly noticed that Daniel was not in the office and so inquired about his whereabouts from Hassan.

Hassan explained, "The poor guy had to leave in a hurry, someone from his house had called and said there was an emergency."

Ethan replied with a sympathetic tone, "Oh my! I hope everything is okay. Well, whenever he comes back can you please ask him to come visit me in my office?"

Hassan agreed and dived right back into his files. After an hour, Daniel came back to the office in sweats as if he ran back to the office. "I hope everything is okay?" asked Hassan. "Yeah, just the usual stuff!" Hassan then informed Daniel that Ethan had asked him to come by.

Daniel quickly made his way to Ethan's office and knocked on the door. "Come in please," said Ethan in a disappointed tone. "I asked you to work on a file a month ago and you haven't given me any updates on that."

Daniel replied, "I am sorry but I got caught up with some family issues, I planned on submitting it, don't worry I will get to it right away!"

Ethan shook his head in discontent and called Hassan to come into the office. "Can you please assist Daniel in submitting this file, I need it urgently and I had asked Daniel to work on this a month ago but he didn't get the time for it." Daniel got disgruntled upon hearing this and intervened, "Why do I have to work with him? I can get it done on my own!"

Hassan tried to calm the situation by saying, "If you need me for anything Daniel, please let me know."

Daniel's aggravation increased and so he said, "I had a family emergency but I have time now, I will finish it on time! I asked you (staring at Hassan heatedly) to inform Ethan that I had to leave due to an emergency! Why didn't you tell him? Or do you not understand the concept of a family emergency given that your family just abandoned you!"

Ethan saw that the situation was getting out of hands and so he said, "Daniel! You don't know what you are talking about, Hassan did inform me and..."

Hassan had heard enough and so he excused himself by saying, "Ethan, thank you for considering me but I think I will pass on this one. If you need me for anything else, I will be in my office."

***

An emotional rollercoaster

"One thing you can't hide - is when you're crippled inside." - John Lennon

Hassan left the office with tears in his eyes which he was fortunately able to hide from his coworkers. Ethan was extremely upset with Daniel but did not want to create a scene in front of Hassan, knowing that it would just make things worse. Since Hassan had left, Ethan continued with Daniel, "What the hell was that? Why did you say such things? Do you even know what happened to his family? He lost them due to an explosion nearby their house which caused the whole house to collapse, they didn't just abandon him, he found him under the wreckage! They were buried alive!"

Daniel was not touched by the story and proceeded, "I know you have a bias for him, you always take his side because he is your friend."

For a moment, Ethan put his head down in sheer dismay and then said, "Yes, he is a friend but I am not here to take sides. We work as a team here and should act like one too! He is a team player and you should become one too. Nonetheless, you have until the end of this week to get the work done and if you fail to do so, you will be facing consequences for your actions today. One more thing, please don't talk about personal issues at work, it is extremely unprofessional and downright unethical, especially the things you said to Hassan, they were very hurtful!"

Daniel apologized and promised to submit the file on time. Meanwhile, Hassan was in his office, quieter than ever. He even skipped lunch with his coworkers. As soon as his work finished, he rushed towards the elevator to head home. Ethan managed to catch Hassan just before leaving. He wanted to make sure if everything was okay with Hassan, hoping that he had not been offended by Daniel's words. Hassan did not want to burden Ethan with his issues and so told him that he was okay.

Hassan left the building and the weather outside perfectly described how Hassan was feeling. The sky was filled with grey clouds with no room for sunshine. It was raining, everybody had their umbrellas out to cover themselves as if they were hiding, the same way Hassan was trying to hide how he felt. Hassan decided to walk in the rain hoping that the rain could wash away how he felt.

As he was walking, he stopped and looked at the sky. It felt like the world was spinning around him, in slow motion. Suddenly, he felt an excruciating pain in his chest as if someone had stabbed him in his heart. He sat down on the foot path to catch his breath.

After a few minutes of resting, he walked to the nearest subway, completely drenched in the rain, and was on his way home. As soon as he arrived, he jumped into the shower. Afterwards, to keep his mind off the things he had been thinking about all day, he engaged himself in cleaning the house.

John had also just arrived and noticed Hassan cleaning the house rigorously. He knew something was wrong and so he asked him. Hassan told him that he was okay but John stopped him from cleaning and saw that Hassan was breathing heavily with sorrow on his face. John understood that Hassan was not comfortable in talking about it and so decided not to antagonize him further. Instead, he hugged Hassan and said, "It's okay, I am here now." Upon hearing this, Hassan could not hold back his tears and so he grabbed John's shirt, put his head on his shoulder and started to weep away.

John tried to stop Hassan from crying but failed to do so. He figured that something must have happened at work since he left in the morning with a positive attitude. He decided to call Ethan and explain what he had witnessed. Ethan promised to come right away. John stayed with Hassan to comfort him not knowing what had happened.

***

A friend in need is a friend indeed!

**"They may forget what you said, but they will never forget how you made them feel." —
Carl W. Buechner**

Ethan came to their house and John decided to take him upstairs to Hassan. Hassan was still crying even though a substantial amount of time had passed.

Ethan said to Hassan, "Please don't cry, I hate seeing you like this." He sat next to Hassan and gave him a warm hug. "You know we are all here for you, share with us whenever you are ready. We are all a family, aren't we? John, Cara and I will always be here for you. Please stop crying for us, you know how much we all love you."

Hassan stopped his tears knowing that he did not want to hurt his loved ones from seeing him in this condition. Ethan asked John to take care of him as he started to depart. John kissed Hassan and asked him to stay in bed until he comes back after a chat with Ethan.

John was able to catch Ethan at the door and asked what happened. Ethan explained that someone at work had brought up Hassan's family in a conversation. John got extremely angry at Ethan and asked him what he did to intervene.

Ethan continued to explain, "I did what I could at the moment, I cannot fight with someone in a professional environment but I did my best. I will make sure that nothing of the sort happens again."

John replied, "Yeah you should be careful with him, especially because you know what he has been through but always remember, I am the one he loves not you!"

Ethan got upset at this and said, "You cannot just cut me from his life like you did before! You loved him just to hurt me, sadly he fell in love with you. Now, if you ever hurt him like you hurt me, I will hunt you down!"

John defensively said, "What makes you think I will hurt him? I love him just as much as he loves me. I learned my lesson from what I did to you but what I have with Hassan is completely different. Wait, don't tell me, you are still in love with him, right?"

Ethan, with tears in his eyes, mumbled, "I loved you! The love I have for him is different. I know I cannot have him now and have made my peace with that. I was very angry when you stole him from me but I have never seen him smile the way he does with you. I just want the two of you to be happy."

Ethan departed leaving a lot on John's mind. All John could think about was how he hurt Ethan so much in the past. Thoughts about Hassan were also haunting him, he couldn't do anything to stop Hassan from crying. The image of Hassan crying was something he despised.

A recurring nightmare

"You learned to run from what you feel, and that's why you have nightmares. To deny is to invite madness. To accept is to control." - Megan Chance

At night, John and Hassan were watching YouTube videos together. Hassan accidently put on an Arabic song which had a very sad tone to it. It was a black and white video which got stuck in John's mind, especially the dance sequence in it. There was a young boy and girl dancing slowly with a glass wall between them. Maybe it represented what John had felt when he wasn't able to stop Hassan from crying. A glass wall in between them not allowing John to come close to Hassan.

Afterward, the two cuddled in their bed and fell asleep. John had another one of his nightmares. This time he saw Hassan lying on a floor and the song that they were listening to was playing in the background. John took a hold of Hassan's hand and started dancing. Hassan started to turn away but John grabbed him and held a knife against his throat. John started pushing Hassan towards a table where he rested him and started kissing him passionately. Hassan began to cry and to stop him, John raised the knife to stab Hassan but woke up to that horrifying image.

He screamed Hassan's name, just as he woke up, and looked around to see Hassan was not in bed. John started to look for him but couldn't find him anywhere in the house. Suddenly, he remembered Hassan's 'man cave'. Hassan would often go down to the basement where he would paint. The basement was very quiet and allowed Hassan's creativity to flow.

John went to the basement where he found Hassan painting. He approached Hassan from the back and hugged him tightly. He whispered into Hassan's ear, "Let's go back to sleep, you know I have trouble sleeping when you're not there to hold me."

***

Paint the pain away!

"Art enables us to find ourselves and lose ourselves at the same time." - Thomas Merton

The next morning, Cara, John and Hassan were having breakfast. Hassan started asking how Cara felt about Kyle. Cara responded, "He is a lovely person to be with. He listens to me and treats me like a lady but I am not looking to fill a void in my life. I know I might sound ungrateful but I do not like the idea of being with someone just so they can help you move on." Hassan assured her that whatever came out of their relationship was her choice and that there was no pressure to be with Kyle if she did not feel like it.

John informed the two that he will be going to Montreal for a few days because of his work. The two wished him the best for his trip and everybody went their own ways. Hassan and John left for work while Cara had to put some things down in the basement. As she was looking around the basement, her attention was caught by an extraordinary painting by Hassan. The painting was of a man with no facial features as it looked like the face was brushed away. The man seemed to be screaming in pain. As Cara looked deeper and deeper into the painting, she was able to see Hassan's suffering in it. Cara was in disbelief as she thought of Hassan being extremely positive and happy all the time. She could now see the burdens he carries with him.

There was a notebook next to the painting which Cara picked up immediately. It read, "Another painful memory that has been locked away, I don't feel like a part of this world anymore. I am stuck in the world of memories and I cannot escape it." Hassan's words had a deep impact on Cara. She felt the same way. However, as she saw the pain consuming Hassan she decided to move on. To be the strength for those around her. She decided there had been enough suffering, pain and dark thoughts. Cara left the basement with a new attitude towards life.

At work, Hassan could feel the pain in his chest again. It was becoming very difficult for him to breathe but he thought it would go away and continued to work. Meanwhile, Ethan was in his office looking at a picture of Hassan and him, thinking about the day they met. Ethan was returning back from his vacation in Turkey. Ethan had asked the flight attendant to change his seat as the person next to him was making a lot of noise. The flight attendant had moved him next to Hassan. Ethan was drinking water when he spilled it all over Hassan. He quickly apologized but Hassan was not offended at all, he smiled and said it was perfectly fine. That smile captured Ethan's heart, like many others. Ethan asked for his contact details and that is how the two became friends.

***

Love is in the air

"You know you're in love when you can't fall asleep because reality is finally better than your dreams." - Dr. Seuss

Kyle wanted to text Cara but was not able to, he felt that Cara was not that into him. He did not want to seem too clingy but he really liked Cara. He was afraid of being rejected as he felt vulnerable exposing himself again to the dating world.

Luckily, he just received a text from Cara. She asked if he was free this evening and invited him to a restaurant. Kyle, without even thinking, said yes instantly. Cara wrote back, "I miss you." Kyle's heart just pumped out of his chest but he took a hold of himself and replied back with, "I miss you too."

Kyle had the biggest smile on his face now, he felt like he had just won a million dollars. He started seeing Cara in everything, the bus driver, the florist, everywhere! Kyle was happy and he wasn't able to get off cloud nine. He didn't want to mess anything up and so called up Hassan for advice. Hassan told him to be himself and let the journey set its pace.

Brian had left some of his stuff at Hassan's house which he went to pick up. Hassan would always leave a key, in case of emergencies, under one of his plants. Cara had told Brian to come and take his stuff. Brian went into the basement as he had never seen it before. He saw the painting and went towards it. Brian, with tears in his eyes, spoke softly, "It's like you're in my mind, reading me. I have a lot of pain in me and you have captured that beautifully!"

Brian picked up his things and grabbed the painting as well. He put the painting in his passenger seat and drove away. The painting spoke to him and said more than words ever could. It sang to him and opened a way to his injured soul. He kept crying, the whole time he drove, looking at the painting. All he could think about was how Hassan had captured him perfectly. He now wanted Hassan more than ever.

Kyle had arrived at the restaurant with a bouquet in his hands. He was wearing a black coat with a red sweater inside. He looked like he came for love. He asked for Cara as he entered the restaurant. A waiter took him to the VIP section. Then he saw Cara standing in a gorgeous blue dress capturing the attention of everything around her, Kyle looked around but there was only one table. Cara had reserved the section for the two of them.

Kyle, finding it difficult to get words out of his mouth, said, "You look breathtaking."

Cara replied whilst blushing, "You look quite charming as well, here is your birthday gift. Happy birthday handsome!"

Kyle was sure that his birthday was a lot happier now. He approached Cara and kissed her gently. Kyle was still unsure about how Cara felt about him and so he asked her.

"You're a sweet guy. I wish more men were like you."

Kyle replied disappointedly, "Oh no! I know what you are going to say next."

Cara denied, "No you don't. You are a real gentleman. I like you and I think we should take more steps to build our relationship."

Kyle now felt relieved and started talking about everything that came to his mind. They were having a good time, laughing and holding hands. Cara couldn't stop smiling. For the first time in years, she opened up to someone romantically. She prevented herself of being exposed to that kind of pain again. However, she was still human and her body and mind desired things which she could only suppress to a limit.

She failed to realize that by restricting herself from a future she was trapping herself in the past. Moving on would not happen overnight but slowly and gradually she was able to make some space in her heart. As she removed the pain from inside, she slowly began to feel joy again. Love penetrated easily. Each daunting memory that she removed from her heart made more space for the love to enter. Kyle was able to make that space in her heart and honestly, they both deserved it. They were held back by their memories for too long and it was time to start their lives all over again, with each other.

Chapter 3: Jealousy

A secret here and a secret there

"There are no secrets that time does not reveal." – Jean Racine

John started packing for his trip to Montreal when he stumbled upon an old wooden box that was locked away. This was not the first time he saw this box. He would always ask Hassan what he kept in inside it, but Hassan would always avoid telling the whole truth by saying things like, "It's just some stuff that I want to keep for myself, nothing important."

John noticed an old notebook beside the box, he picked it up and saw a poem on one of the pages, which read:

"I still remember the day I said goodbye to you

Unbearable pain beyond all limits

I thought I could forget about you

It wasn't possible because I still miss you

I still want to see you fall asleep in my arms

But that only a beautiful dream like it was

Now, nothing but a memory and pain

I will always be a slave to that memory

The pain was born on the day you died."

John kept reading, but some of it was written in Arabic and so he decided to put it back in its place. He always wondered what Hassan was hiding from him, but now he was devastated to find out that Hassan was not truly his. Hassan was still living with a broken heart, and John felt like maybe Hassan was never able to fall in love with him due to his past. He started feeling the emptiness of the room resemble the emptiness that began to form inside him.

He was staring into the abyss, talking to himself, "Was it all a lie? No matter what I do, he will always be in Hassan's heart." John turned to look at the time but saw Hassan standing at the door, "You're mistaken; you are the only one I am in love with." John was not satisfied with the answer and so, he started asking about the box and the notebook. Hassan was not comfortable talking about it but he asked John to check the date on the notebook. The date was before Hassan had even met John. Hassan said with tears in his eyes, "Don't make a fool of yourself. You know I am deeply in love with you."

John quickly realized his mistake and hugged Hassan. He noticed that Hassan's temperature was a bit high and he was breathing heavily. John wanted to take Hassan to the hospital but he refused claiming that it will be okay after he gets some sleep. John continued his packing while Hassan was able to catch a short nap.

Hassan woke up and found John watching tv downstairs. John had ordered pizza and the two sat together to enjoy their meal. "I am sorry for what I did today, it's just that I love you to the extent that it overtakes my heart and consumes my soul," said John. Hassan reassured him that he wasn't angry at him and that he realizes the emotional baggage that he has burdened John with.

They ate their pizza while watching one of their favorite shows 'Hot in Cleveland'. The two laughed until Hassan fell asleep on John's chest while he was holding him from behind. John loved watching Hassan sleep on him but he noticed that Hassan started breathing heavily again and his body was twitching. "Another dream," John thought. He picked Hassan up in his arms and took him to the bedroom. The two fell asleep holding each other tightly.

***

John sees himself in a school playground holding a camera. He tries taking some pictures but is unable to. He notices that he is barefoot and it is raining. There are no water droplets but small pieces of pictures. He looks up to the sky and the small pieces of pictures start burning to ash. The ash starts covering the whole playground and he then sees a butterfly with burning wings. Suddenly, he notices a man on the ground crying. John asks the man if he is okay, instead of replying he just hands him a picture. It is a black and white picture of a man, screaming. He looks at the man but he starts fading away just as the rest of the playground. John panics and closes his eyes and wakes up. Hassan hugs him and says, "It was a bad dream but the real nightmare starts now!" stabbing John with a knife. This is when John actually woke up, horrified.

'These nightmares are becoming frequent now' he thought to himself, then moved closer to Hassan and tried falling back to sleep.

A trip to the hospital

"It's healthy to be sick sometimes." - Henry David Thoreau

In the morning, John and Hassan were having breakfast while Cara wasn't home. Hassan hadn't said a word since he woke up, so, John asked if he was alright. "I am okay, just feeling a bit dizzy." Hassan did not want to trouble John as he knew that he had an important meeting today. John rushed to work and Hassan had to get an Uber. On his way to work, Hassan felt a throbbing pain in his chest and it was becoming extremely difficult to breathe. He managed to make it to work but Ethan instantly noticed that something was wrong. He asked Hassan if he was alright.

Hassan replied weakly, "I cannot breathe."

Ethan replied, "Let me take you to the hospital, should I Call John?"

Hassan did not want to upset John on his big day, so he asked Ethan not to tell him. He rushed Hassan to the hospital and hurriedly began to check him in. Hassan's condition was worsening by the minute and it was taking a long time to get him a room. Ethan could see that Hassan was trying to hide his pain but it was just too unbearable for him. "I wish I could take away your pain," Ethan said to himself. After waiting for two hours, they were finally able to find a room.

A physician entered and said, "Hi, I am Dr. Andrew and this is Nicole, the nurse who will be with us. Hassan, can you describe the pain that you are having?"

After getting some information from Hassan and his blood sample, Dr. Andrew consulted the attending doctor. The two discussed his condition and believed that it might be pneumonia. However, they needed an X-Ray and a CT scan just to be sure.

Dr. Andrew informed Hassan, "They will be taking you for your X-Ray and CT scan shortly, how is your pain now?"

To this, Hassan replied, "I'm feeling much better, the nurse gave me a painkiller. By the way, you are adorable and my friend here is single."

Ethan embarrassingly interrupted, "Excuse my friend; he didn't mean what he said."

Hassan joked, "Look at him, he is so cute."

Ethan surprisingly asked, "What have you given him? He is never so flirty or humorous in such a situation."

Andrew explained that since this was the first time Hassan had taken a painkiller, it could be influencing him. However, Hassan kept making comments to set Ethan up with Andrew. Andrew ended the conversation by telling Hassan that he is into short guys and left the room.

Ethan said, "He was into you!"

Hassan replied, "Yeah, but I am in a serious relationship."

Ethan joked, "It isn't written on your forehead. Plus, you are so different from the other guys, anyone can fall for you."

After the tests were conducted, Dr. Andrew came into the room with the results. "Hassan, you have symptomatic subsegmental PE that requires only short-term anticoagulation for three months," explained the doctor. Hassan and Ethan did not understand the scientific terms and so Andrew had to break it down for them, "You have two clots in your right lung, and the pain you experience is the dying tissue."

Andrew scheduled a follow-up appointment with a Hematologist and prescribed him Rivaroxaban, an anticoagulant medication used to treat and prevent blood clots. As Andrew was leaving the room, he asked Ethan to speak to him outside. Hassan had fallen asleep by the time.

"I want you to take care of him," said Andrew, thinking that the two of them had a special bond. Ethan explained that Hassan is already in a relationship.

"I can see he is very special for you. At first, I was jealous, I thought you were his partner," said Andrew.

"Years ago, I had the chance but ended up losing him to another guy," replied Ethan.

Andrew was interested in finding out more, so he asked Ethan to meet over a cup of coffee and they exchanged their phone numbers.

***

Love is a remedy for all

"Treat a sick man with medicine and a sad man with music." - Amit Kalantri

Ethan called John to tell him about Hassan's condition. John was busy in his meeting and was not able to come. He asked Ethan to take care of him till he gets there. Hassan had woken up by now and was ready to be discharged from the hospital. Ethan took him home and made soup for him. "You are spoiling me," said Hassan with a smile on his face.

"You enjoy the soup, I will go get your medication till then, by the way, where is Cara?" asked Ethan.

"Kyle and Cara have taken their relationship further. I am so happy for them, they are spending more time with each other," replied Hassan.

Hassan starts asking Ethan about Andrew and when he finds out that Ethan was able to get Andrew's number, he couldn't stop teasing him.

"I am in love with someone else," said Ethan.

"I know he broke your heart," whispered Hassan.

"No, he didn't. I never told him how I feel about him," said Ethan dramatically.

Ethan came back from the pharmacy and he ran into John at the door. John thanked Ethan for being there and asked about all the details. Ethan informed him that Hassan had two clots in his right lung. Upon hearing this, and the pain Hassan must be going through, John rushed inside to hug his loved one. He decided to cancel his trip but Hassan reassured him that Cara and Ethan would be there to take care of him.

Ethan was on his way home when his phone started to ring, it was Cara. She had called to find out how Hassan was doing, as John and Hassan didn't tell her much and asked her to stay with Kyle and focus on their relationship for now. He said the same things to comfort her and told her he'd take care of Hassan, so she doesn't have anything to worry about.

***

Hassan reeked of the antiseptic chemicals used in hospitals and decided to take a shower with John. John stripped Hassan and checked him out from top to bottom. He bit his lip and whispered to Hassan, "I want to eat you."

Meanwhile, Kyle and Cara were dancing surrounded by dimly lit candles. Kyle idealized her as she had everything he ever wanted in his partner. Cara felt alive and happy at the moment and wanted to open her heart to him.

Every relationship is like a dance. In the beginning, when the song is slow, it is easy to follow each other's steps. However, as soon as the rhythm changes and one person makes a wrong move, then the whole dance is disrupted. The same can be applied to a relationship. When things are going smoothly, it is very easy to manage the relationship but one slightly wrong move can shake the foundations of it.

Everything was going smoothly for Cara and Kyle. They had managed to come close in a short time as they shared their sorrows and had opened up to each other about their pasts. They had started getting more physical. Cara felt electricity run through her body whenever Kyle kissed her. She thought she had forgotten this feeling years ago and that no one would ever be able to make her happy, but now, the butterflies had once again invaded her stomach.

That night, everyone had fallen asleep peacefully and with their loved ones. John was holding Hassan, while Cara fell asleep in Kyle's arms; this was the first time in years when he slept with a smile on his face. Brian fell asleep holding on to Hassan's painting. The only person who was struggling to sleep was Ethan. He looked at the empty bed space beside him and hoped that someday someone would be there.

***

Revenge is a dish, best served cold!

"Not forgiving is like drinking rat poison and then waiting for the rat to die." - Anne Lamott

The next morning, John prepared for his flight to Montreal. He asked Hassan to take care of himself and update him every now and then. Hassan had taken his medication and was feeling dozy. John tucked him into bed and kissed him goodbye. Fifteen minutes after John had left, Brian came over to the house and opened the door using the key that Hassan kept under the plants.

Brian went to the room he was staying in to grab some of his things. While making his way there, he noticed that Hassan's door was left open. He entered and saw Hassan lying in bed, and there were some medications on his nightstand. He picked them up and read that anyone who takes these will sleep like a rock. Knowing this, he removed the blanket on top of Hassan. He was completely naked under the sheets and Brian said to himself, "Wow you are hairy, just the kind of man I want." Brian knew he had the perfect setting for his revenge that he had been planning for some time. He took out his phone and started taking pictures of Hassan, naked in bed. Then he proceeded to feel and kiss, all over Hassan's body.

Brian left, but he forgot to lock the door. He drove back home, thinking about Hassan's naked body throughout the way. Hassan woke up to a call from Ethan, who just wanted to make sure that Hassan was okay and told him that he would be visiting him later. As soon as Hassan hung up, he got a call from Cara. She was calling to check up on him. Cara also wanted to make some changes to the menu of their restaurant, Hassan assured her that she didn't have to ask him for anything.

Hassan wanted to do something productive instead of staying in bed all day, and so, he decided to rearrange the room. As he was going through all of the stuff in their room, he stumbled upon an old photo album. He opened it as he had never seen it before. He couldn't stop smiling as the pictures were of John from his time in school. Suddenly, the smile vanished. He saw a picture of Ethan and John kissing. He had asked John repeatedly over the years if he knew Ethan from before, and he always denied it. Hassan was extremely confused but when he flipped the picture, there was something written. His tears started to flow like a waterfall. He was shocked and was unable to move. It was like time had stopped for him. He had found out that John and Ethan were engaged! He put the album back where he found it and resumed his rearrangement with an awfully heavy heart.

Brian had taken the pictures but was unsure what to do with them. He used an app to hide his caller ID to send the photos to John. "I want you to burn in jealousy and doubt, which will only lead to darkness," Brian thought to himself.

John received the photos and got so angry that he smashed his phone against the wall. His breathing got heavy as he asked himself why this happened. He stormed out and ended up in a gay bar. He was feeling lost and wanted something to distract him from reality. He started flirting with anyone he could find. He was trying to find ways to hurt Hassan as much as he hurt him. A pair of guys had started talking to him, and they were overtly sexual. They invited John to come over to their place for a fun night; he agreed and went to their place.

They immediately started getting physical with John on the couch. He didn't want to kiss anyone, so he unzipped his pants and started touching himself. Ten minutes later, they were completely naked on the bed, grabbing each other of what can only be described as one of the darkest sexual experiences he has ever had.

Even though John was with them at the time, all he could think about was Hassan. He was hurt since the man he loved was cheating on him. The confusion and darkness were killing John slowly. He even dropped a few tears. He kept visualizing Hassan with another man, laughing at John while kissing each other. John left that place in the morning. As soon as he was out of the building, he broke down crying.

***

Facing the truth

"Truth without love is brutality, and love without truth is hypocrisy." - Warren Wiersbe

Hassan was crying over the pictures he had just discovered. The doorbell rang but he knew it would be Ethan and he was not interested in talking to him right now. Ethan tried calling Hassan but he wouldn't pick up. Ethan knew about the spare key, and so he opened the door as he was worried about Hassan's health.

Hassan shouted, "I want to be alone!"

"You are avoiding me. What happened? What did I do to upset you?" asked Ethan.

Hassan angrily replied, "Well, if you are that concerned, why don't you tell me the story of you and John? Were you both engaged?"

Ethan was ashamed now that Hassan had discovered their secret. He couldn't lie anymore and resorted to saying yes, as lightly as possible. Hassan wanted to know why he never told him all this before. Ethan explained that he was in a very difficult situation when Hassan chose to be with John. He told him that he loved him from the start and didn't want it to seem like he was trying to cause fights between the two for his personal gain.

"You preferred not telling this to me for years. Don't you think I deserve the truth?" asked Hassan.

"Do you really want to know? Even if it hurts?" asked Ethan. Hassan replied loudly, "The whole truth!"

"Okay, I will tell you the short version. John and I loved each other a lot until we got engaged. My control and obsession, coupled with his jealousy, destroyed our relationship. When I went to Turkey, I was in a very bad place because of what happened. When I met you on the plane, I fell in love with you. It was love at first sight. When John figured out my feelings for you, he decided to get his revenge by stealing you away, but instead he fell in love with you during all this," explained Ethan

Hassan felt betrayed by both of them, but Ethan explained that the smile that John would bring on Hassan's face would not allow him to intervene. Ethan wanted the best for both of them.

"I was into you too! I was waiting for you to make a move. Now, I am in a serious relationship based on what?" said Hassan.

Ethan kept apologizing but Hassan was in no mood for forgiveness. All he could think about was how they lied to him, played with his feelings, and used him for their games. Ethan explained that he never used him and that is why he chose to stay away from all this. He promised Hassan to always be by his side.

"If John and I ever break up, I can never do what he did to you and so, I can never be with you," said Hassan.

Ethan wasn't worried about that, he just wanted Hassan to be happy. "All of this must've been very hard for you," asked Hassan. "Don't worry about me, I can handle it and I won't make such a mistake again," assured Ethan.

Hassan fell asleep in Ethan's arms after a while. Ethan knew that Hassan was probably very tired, but he couldn't stop himself from loving the way Hassan was sleeping in his arms. Even though he can't love him openly, he enjoyed that moment to the fullest. He kept staring at Hassan, observing each breath that he took. He moved closer to him and closed his eyes. Hassan's phone started ringing but Ethan put it on silent as he fell asleep holding Hassan in his arms.

Chapter 4: The Monster

Consequences of a truth

"Rather than love, than money, than fame, give me truth." - Henry David Thoreau

After a couple of hours, Kyle and Cara had returned home to find Hassan asleep in Ethan's arms. Cara was shocked and thought to herself, "Is Hassan cheating on John?" Ethan woke up and saw the disbelief on Cara's face. He immediately put Hassan on the couch and greeted Cara.

"What is going on?" Cara asked.

"We had a long conversation and he was crying a lot, I held him to comfort him and we fell asleep," Ethan explained.

"Why was he crying?" asked Cara.

"I told Hassan about my past with John," Ethan answered.

"What past?" Cara asked as this was all new to her.

"We were engaged a long time ago, we hid this from everyone, including Hassan, but he found out. I am not a bad guy for hiding that, right?" Ethan asked.

"How are you not the bad guy?" asked Cara.

"I loved Hassan and did not want to hurt him; I always saw what John and him had, and I could not come in between that. I thought John would tell him eventually, but he did not," explained Ethan.

"Tell me everything from the start. Please do not try to hide anything this time," said Cara.

Ethan began to tell his version of the whole story. Cara and Kyle listened carefully; she was worried about the pain that Hassan must be going through. Hassan woke up and approached the group but they asked him to go back to sleep.

After a while, John called and Cara was the one who answered. John wanted to inform Hassan that he would be staying a few more days because of his work. Cara informed him that

Hassan was asleep and that he has not been well since the morning. John started worrying about his health but Cara told him that everything was under control.

Ethan wanted to leave because of all the embarrassment but Cara asked him to stay, to take care of Hassan. She told him that he could borrow some of John's clothes and Ethan agreed to stay. He went to Hassan's room and knocked on the door.

"Come in," shouted Hassan.

Ethan asked, "Would you mind if I stayed the night with you?"

Hassan replied, "I would like to! Cara told me you wanted some clothes but I don't think anything here suits you. You can just keep your underwear on."

"I don't think I will be comfortable in that," replied Ethan.

"Nothing will happen between us. I trust you and you know what, I will also roam in my underwear to make you feel comfortable," said Hassan in a reassuring tone.

"How can you trust me? You know that I like you, I have always wanted to cuddle and sleep with you for, at least, a night," said Ethan.

Hassan took off his pajamas and laid on the bed. Ethan followed him into bed and held him tightly in his arms. Hassan asked Ethan if he loved him, Ethan looked deep into his eyes and kissed him, "You are driving me crazy," said Ethan. They started feeling each other's body at which Ethan stopped to ask, "What about John?" asked Ethan. "I am with you now," replied Hassan. As they continued, Hassan yelled, "Oh yeah!" Ethan woke up and found himself wet, sleeping next to Hassan. Hassan was still asleep in his pajamas and Ethan wearing his clothes; and thought to himself, "It was just a dream."

During this time, Cara and Kyle had gone to bed. They were discussing what Hassan was suffering through. Kyle assured Cara that Hassan would be alright since he had her to take care of him. Cara was upset but wanted to enjoy her time with Kyle. She held his hand and asked if he wanted to dance. Kyle was ecstatic and he quickly played 'sway' by Frank Sinatra. Kyle grabbed her by the waist and started to dance to the song.

"If you go away

On this summer day

Then you might as well

Take the sun away

All the birds that flew

In the summer sky

When our love was new

And our hearts were high

And the day was young

And the night was long

And the moon stood still

For the night bird's song"

As they danced, Kyle closed his eyes to feel the moment. All he could see was her glowing face. He leaned towards her and paused just before he could feel her lips. He looked at her, but her eyes were closed, that was all the permission he needed, and so, their lips met. At first, he could feel her tenderness but then she grabbed him by his hair and pulled him towards herself, passionately. She opened her mouth just enough for their tongues to interlock. Kyle picked her up and carried her towards the bed. They started undressing each other and the night took a wild turn. They had found their connection, and they were fervent about each other. Their bodies knew that they were meant for each other.

***

A week had passed, and John had arrived home. The house was filled with silence throughout. No one had anything to say to each other, not even a single word. This was probably the most awkward moment for this household. Hassan was irritated by the awkwardness and so, he excused himself by going downstairs to the basement. Cara told him that she would bring him a snack.

"You are not going to say anything to him?" asked Cara in disbelief.

"What do you want me to say?" asked John.

"You don't have to say anything, he is trying to move on. Kiss him, make love to him. I don't know, but do something! Don't you miss him?" asked Cara.

"Of course, I do, but I think he wants to be alone right now," explained John.

"No, he doesn't, he needs you right now," explained Cara.

John replied, "You only care about him. What about what I need?"

Cara replied, "I want him to be happy and you are his happiness. I care about both of you but if I had to choose, then he would always come first."

John murmured, "Nobody cares about me."

Care explained, "That is not true at all! Hassan cares a lot about you and he loves you. Can't you see that?"

John replied, "You are right, I love him too!"

Cara replied, "Then go show it to each other! Stay happy, I care about both of you."

John rushed downstairs and sat behind Hassan. He smelled the back of Hassan's neck and remembered all of their good times. He started to lick Hassan's neck as Hassan closed his eyes. John positioned Hassan's legs around his own waist. Hassan was still hurt and John could see that but it seemed to turn him on even more. He started kissing Hassan roughly as if he was a lion pouncing on its prey. After a while, John stopped to cuddle Hassan.

"I am sorry for being rough," said John.

"I don't mind anything as long as I am with you," replied Hassan.

John's eyes turned to a painting that had not been completed by Hassan. He asked Hassan if that was him in the picture, as the face resembled John's. Hassan confirmed that it was a painting of John.

"But I don't have any pictures like this, how did you draw me so perfectly?" asked John.

"I don't need any pictures to draw the love of my life," Hassan replied.

"I love you, shorty," said John.

"This is the John I missed; I love you too!" said Hassan.

"Are you up for some more action," asked John while biting his lower lip.

"You don't need to ask," replied Hassan.

"I might add more pain to our pleasure. I know you don't like it when I get rough but what can I do? I just want you so badly," explained John.

The two continued making love to each other and things seemed to be heading in the right direction for them. Some time had passed, Cara wanted to know what the boys were up to and so, she shouted from behind the door, "Are you love birds done? I have made snacks for you guys, please come upstairs when you are ready." John and Hassan went upstairs immediately, as all that action had made them very hungry.

"You both are positively glowing, good job kids, you made mama proud!" said Cara in joy.

John replied, "Thank you ma'am, can we eat now? I am famished!"

Cara joked, "Haven't you already had your dessert?

John replied, "Look, you made him blush. Isn't he adorable when he blushes?"

Hassan intervened, "Stop it, you two!"

Cara could see that Hassan was much happier now, she smiled as she felt relieved that her best friend is now okay. John informed them that he will be visiting his mother and that they should not cook anything as he will bring food on his way back. "Say hi to her from me and give her a kiss on the forehead," said Hassan, as John headed towards the door.

***

A lost soul

"Running away was easy; not knowing what to do next was the hard part." - Glenda Millard

On his way back home, John stopped at a convenience store to get some beverages. Brian was already at the store as if he had planned to be there.

"Hi John, how are you?" asked Brian.

"I am alright, thank you for asking. How about yourself?" asked John.

"I am well. Look, I wanted to apologize for my behavior on the first day when I moved in with you guys," said Brian.

"What are you apologizing for?" asked John.

Brian replied, "I was drunk when I found you sleeping in your room. I was naked when I was approaching you but Hassan came and handled the situation really well. I am sorry but I was drunk, I don't know what happened to me that day."

"What are you talking about?" asked John.

"Oh. I am sorry, I thought Hassan already told you all this," said Brian.

"What happened exactly?" John asked as he was getting impatient by the second.

"Well, I was drunk..." said Brian but John interrupted him, "Yeah, you made that very clear, tell me what did you do?"

"I tried to have sex with both of you but, Thank God, Hassan stopped me," explained Brian.

"Get out of my face before I do something stupid to you!" yelled John in extreme frustration.

Brian apologized again, and walked away, however, he had a sinister smile on his face as if he had just made a deal with the devil. He knew he was successful in getting John angry. John had started breathing heavily as he hastily got in his car. He drove quite rashly while his mind kept filling with negative thoughts. He reached home and bashed through the front door.

John screamed at the top of his lungs, "HASSAN! Come here right now!"

Hassan hurried to the door; Cara also followed behind.

"What happened? Are you okay?" asked Hassan.

"You are such a liar and manipulator; how could you lie to me?" asked John.

"What are you talking about?" asked Hassan.

"Why didn't you tell me what happened between you and Brian, on the first day he moved?" asked John.

"Cara, please go to your room," asked Hassan as he did not want to create a scene in front of her.

"Why? She needs to know how you manipulate people!" exclaimed John in anger.

Hassan, now getting angry, yelled, "Shut up! You have no right to call me a liar or manipulator!"

"Then why didn't you tell me? What else are you hiding from me?" asked John.

"As far as I remember, I told you what happened the next day, the only thing that I didn't tell you was that it was Brian," said Hassan.

"Flipping the truth, as always. I am sick and tired of you using me and my feelings," said John.

"You have no right to call me names! I should be calling you manipulative, you lied to me from the start of our relationship," said Hassan.

"What are you talking about?" asked John.

"I am talking about you and Ethan," said Hassan in disappointment.

"What? H-h-how do you know about this? I will kill Ethan!" yelled John in anger.

Hassan, with tears in his eyes, explained, "He didn't tell me anything. I found your photo album. Did you ever love me for real?"

John took a deep breath and replied, "I know what I did was extremely stupid but I genuinely fell in love with you."

"You just used me to get your revenge on your ex!" Hassan cried.

"I love you," John said, trying to defuse the situation.

"You lied to me," accused Hassan.

"You lied to me too!" said John.

Hassan replied, "When did I ever lie to you?"

John had to think of something other than blaming Hassan for cheating. The reason being, if they both cheated on each other, John was afraid of losing Hassan forever. He was thinking of what to say. "Answer me!" shouted Hassan. John had to say something so he used the wooden box, "You lied about your secret box, didn't you?" Hassan looked at John with disgust and then rushed upstairs to get the wooden box.

Hassan opened the box and asked, "You want to know what is inside this box, right?"

Cara intervened, "Hassan, please calm down."

Hassan was crying as he pulled out a half-burnt picture from the box. "This picture is of my sister, I found it under the wreck after the explosion destroyed my house." Then Hassan pulled out a broken pocket watch, "This was my father's watch." Then he took out a journal which was still ashy from the burns, "This is my mother's journal." Then he took out a bracelet, "This was my brother's." Then he took out a watch, "This was my first love's watch. I took it after the militia slaughtered him. These are the only memories that I have of my family, the loved ones I buried with my own hands. This is all I have left of them."

"I am so sorry, I had no idea," explained John.

Hassan replied with a weak voice, "Are you happy now? I never lied to you. I just wanted to keep my darkness to myself but I clearly failed."

Hassan gathered all of his memories and placed them back inside the box. He went to the basement to put away the box. John tried to follow him but Cara stopped John from leaving.

"You have done enough," said Cara.

"But..."

Cara interrupted John, "Leave him alone! You have already done enough damage. I cannot believe you sometimes. How could you do this to him?"

"I am an idiot, I know. I got carried away," explained John.

"I hope you are happy now," said Cara.

"But I love him," said John.

"I don't believe you anymore," said Cara, highly disappointed with John.

Cara went to the basement to talk to Hassan but she found out that he was not there. She immediately shouted for John. John ran downstairs and they realized that Hassan had left a while ago. They ran outside to look for him but it was raining heavily and Hassan was nowhere in sight. John started blaming Cara for not letting him follow Hassan. "You have already done enough, the reason he left is you! We need to find him first, then we can talk about all this," explained Cara.

Cara called Ethan to ask for his help in finding Hassan. Ethan was on a date with Andrew, the doctor from the hospital he and Hassan visited. Cara asked for his help and explained that Hassan left in a disturbed state. Ethan told her that he knows where he would find Hassan.

"I am so sorry but this date is over. I need to go, it is an emergency," explained Ethan.

"Is everything okay?" asked Andrew.

"Hassan is missing and we are all trying to look for him. I am so sorry that I have to go," said Ethan.

Andrew insisted on going with Ethan to look for Hassan. "I am sorry for ruining our first date," said Ethan. "Don't worry, you will get many chances in the future to make it up," said Andrew. Ethan and Andrew got in the car to begin their search. Ethan was driving like a maniac and Andrew kept trying to calm him down. Shortly, they arrived at The Lake Ontario Park. Andrew asked what they were going to do there. Ethan assured him that they would find Hassan here.

They found him sitting on a bench, holding his box and he was talking to someone. It seemed like Hassan had been sitting there for a while as he was completely drenched in the rain. Ethan got a closer look and could not believe what he saw. Andrew asked Ethan who Hassan was talking to. Ethan said he would explain it to him later and that he should wait for him to deal with Hassan.

Ethan sat beside Hassan and moved his umbrella over Hassan's head. Ethan had never seen him like this, it was heartbreaking. Ethan knew that Hassan felt shattered. He only wondered what had happened to Hassan.

"It's time to go home Hassan," said Ethan.

"I don't want to go home," replied Hassan.

"You can come over to my place," explained Ethan.

Hassan turned to his friend and said, "Take me with you, I don't want to be here anymore."

Ethan slapped Hassan's face and yelled, "You are not going anywhere! Do you hear me? You are not going with him!"

Ethan dragged Hassan to the car and wrapped him in a towel as he put Hassan on the back seat. Ethan informed Cara that he would be taking Hassan, to his place, for a few days and that he will come by to pick up Hassan's things from their house. Andrew asked Ethan if they were going to his place, to which Ethan replied yes.

"I am sorry for the trouble you guys faced because of me," said Hassan.

Andrew replied, "Don't worry, we are alright."

Ethan jumped in quickly to say, "I want to be sure that you are alright Hassan, you will be staying with me for a few days."

"I don't want to bother you," Hassan replied.

"You can never, you are my family, don't ever forget that," said Ethan.

"I know that, I don't know how to thank you," said Hassan.

"We will go pick up your things first, if you need anything in particular, let me know," said Ethan.

As they went to Hassan's house, Ethan kept checking his rearview to keep an eye on him. He was so tensed about Hassan; he had forgotten all about Andrew. Andrew had figured that Ethan has feelings for Hassan but he was confused because Ethan always called him 'family.' Andrew was having a hard time understanding their relationship and he didn't want to be in the middle of something so upsetting. Hassan had fell asleep during the ride and Ethan noticed that he was shivering.

"You're in love with him, aren't you?" asked Andrew.

"Yes, is it that obvious? It's a long story, but Hassan and I, will never be together," explained Ethan.

"What do you mean?" asked Andrew.

"Hassan and I, can only be family or friends, nothing more than that," explained Ethan.

"I understand," said Andrew.

"If you don't want to see me again, that makes sense," explained Ethan.

"Well, I can never compete with him," said Andrew.

"You could be more if you want to," explained Ethan.

"No one knows what the future holds," said Andrew.

They had arrived at Hassan's house and so, Ethan went inside to get his things. Cara asked Ethan about Hassan's condition. He informed her that he is asleep in the car, although he was completely drenched in the rain when Ethan found him. John had also joined in on the conversation.

"Where did you find him," asked John.

"I am not talking to you, I feel like punching you in the face so hard," said Ethan.

"Go for it," dared John.

"Both of you shut up! John, please don't aggravate the situation further," said Cara.

"This is between my boyfriend and I; Ethan has no right to interfere in all this," explained John.

"You have involved me in all of this by keeping our past a secret from him! Hassan will never do what you did to me. He is a much better person than you will ever be! Anyway, there is no point in arguing, please give me his clothes and medicines, he will be staying with me for a while," explained Ethan.

"I cannot allow this," said John.

Ethan explained, "He doesn't want to be with you right now, please give him some space, he really needs it."

John understood that Ethan was right and so, he went to grab Hassan's things. Cara apologized to Ethan for asking help. Ethan was only interested in finding out what had happened. Cara promised to tell him at a better time.

"I cannot imagine what he has been through, he looked so fragile when I found him, he even had his box of memories with him," explained Ethan.

"Do you know what is inside that box?" asked Cara.

"Yes, he showed it to me when my mother passed away. He was the only one who helped me during that phase," explained Ethan.

"You do understand what you and John did to Hassan was very wrong?" asked Cara.

"I know, I tried to tell him but whenever I see him with John, the smile on Hassan's face is invaluable. I don't want to be the person who takes away that smile," explained Ethan.

"It wasn't the right choice," explained Cara.

Ethan agreed but it was too late now. John brought Hassan's things and handed them over to Ethan. Ethan thanked him, said his goodbyes and departed for his car. He switched the ignition on, started driving and headed to his place with Hassan asleep in his backseat.

Chapter 5: The Chef

When will these nightmares end?

"Which is the true nightmare, the horrific dream that you have in your sleep or the dissatisfied reality that awaits you when you awake?" - Justin Alcala

John was still upset over Hassan leaving and the way Ethan and Cara had talked to him. John confronted Cara by saying, "I would mind my own business if I were you. I am his boyfriend while you are a nobody. I hope you can find a place to move in." Cara was not to be bothered by John's words as she replied promptly, "Don't you dare call me a nobody. I am a part of his life. I will stay here as long as Hassan wants me here." Cara walked away as she said that to John. Meanwhile, John kept thinking about how Hassan keeps playing the victim and gets away with everything.

Ethan had taken Hassan to his place after dropping Andrew at his place. As they entered Ethan's apartment, Ethan asked Hassan to take a warm shower to help ease the cold. Hassan didn't seem to have the energy for it, and so, Ethan helped him shower by taking off his clothes and standing with him, just in case he needed anything. This was the first time Ethan had seen Hassan completely naked; Ethan felt aroused and found it difficult to control himself. Nonetheless, he succeeded in suppressing his feelings and put Hassan to bed and fell asleep beside him.

Ethan woke up as Hassan kept saying that he couldn't breathe. He decided to call 911, but Hassan warned him that no matter how much he tried, he will not be able to save him. As Ethan grabbed the phone, it turned into water. He looked back at Hassan and he was held captive by John, holding a knife to his neck. Ethan begged John to stop but he dismissed his plea by saying, "You and Hassan will never be a thing." Suddenly, everything around Ethan disappeared until he was left in a black void. Ethan could hear someone breathe in all the darkness, he called out to Hassan, thinking it's him. Hassan responded to Ethan and so, he tried to find Hassan by following his voice. As he found Hassan, he could only see half of his face with snakes covering all of his body. Hassan smiled mockingly as he said, "I told you, you can't save me!" Ethan woke up frightened and realized that it was a dream.

Back at Hassan's place, John was also having trouble sleeping. He kept thinking how his life changed overnight, from having a lover to a cheater. He kept staring at the ceiling, trying to find meaning in his life. He suddenly thought of Hassan's painting and went over to look at it. He was remembering Hassan through his painting when his eyes landed on Hassan's notebook; he opened the last page to read:

"He lights up my life. I would burn the heart of mine into flames for him, and the ashes of my heart will be left behind. Where ever I look, he is always there. I cannot describe the feeling, but I have a name. John, your name will forever be engraved in my heart."

John stopped reading and took a deep breath. He was mesmerized by Hassan's words and hugged the notebook close to his heart. He laid back on the couch and closed his eyes. Suddenly, he heard Hassan asking him to open his eyes, as he does, he finds himself in another place. Hassan tells him that it is his secret place. There was a waterfall there and both of them held each other's hands to walk towards it. They started playing with each other in the water. The playing quickly turned into a romance between the two as they started kissing each other, under the water. Unexpectedly, everything around John turned into a similar void, like that in Ethan's nightmare. He saw himself chained up and the ground underneath started crumbling. As the ground split apart, he started falling to no end. There was only darkness, but then a voice prevailed. It was John's own voice talking to him. It warned him that he was nothing but destruction. Abruptly, John stopped falling and was now floating in the darkness. He was unable to move and saw something approaching towards him very quickly. John screamed, "No," and woke up.

In the morning, Cara called Hassan to check on him. Hassan told her that he was fine, but he could tell that something was wrong with Cara. Cara asked Hassan if he wanted her in the house. Hassan immediately understood that John probably caused all this and so, he said, "This is your home; no one can ask you to leave. I want you to stay there until you have kids so that I can be their cool uncle." Cara ended the call on that funny note and was now feeling much better.

***

Ethan was driving Hassan to work when John called. He wanted to apologize to Hassan but he told him that he still needed some 'alone time'. Hassan also warned him that this was his second mistake. John understood and just wanted to make up with Hassan. Hassan allowed John to take him out on a date. Hassan ended the call, but Ethan was upset. Hassan explained that he had so much love for John that he had no choice.

In the evening, John took Hassan to a Korean restaurant. The dinner was served and chopsticks were the main utensils to be used. Hassan was embarrassed as he had never used them before. John told him to leave the chopsticks and that he would feed Hassan. Hassan felt loved again and he said, "I love you, please don't hurt me anymore, I can't live without you."

After dinner, they decided to get ice-cream. John asked Hassan to put all the flavors on his lips so that he could taste them. One by one, John enjoyed all the flavors and complimented Hassan's lips for the added flavor. They were near Christie Pits Park and ended up on a bench there, with John sharing silly stories to make Hassan laugh. John asked if Hassan would come home tonight. Hassan agreed and John enthusiastically picked him up to take him to the car.

As soon as they got home, they started undressing each other. They were touching, kissing and making love when Ethan's call interrupted them. John picked up and dismissively told him to call back later as they were making love at the moment. Ethan saw Hassan's bathrobe lying on the floor; he hugged it tightly and started smelling it. He started feeling the bathrobe and then he suddenly stopped, realizing how stupid he was being. He cried to himself, "Why can't I have you?"

As the heat died down between Hassan and John, Hassan reminded John that he was one mistake away from destroying their relationship. John's eyes started to tear up as he asked Hassan why he was being so harsh with him. Hassan replied, "You used me for your games with Ethan and then you called me manipulative. If we cannot trust each other, how will we last?" John went on to ask if he had any feelings for Ethan, given that he was the only person who knew where to find Hassan. Hassan explained that they used to go there together and cleared that there was no such thing between them.

John praised Hassan for what he wrote about him and both of them fell asleep holding each other. John saw himself in his parent's house and his father was holding a belt in his hand.

"You are late again," said John's father.

"Why are you here?" asked John.

"Did that fag teach you to disrespect your father? Let me teach you what respect is," his father said.

He got up and started hitting John with his belt. John was unable to move and his father kept hitting him. John's mother, Marguerite, interrupted, "I won't let you hurt my son anymore. It would be best if you left us alone. We don't need you". She held her son tightly and kissed him on the head.

"I don't want to see him again; do you think I'd be better off if I was straight?" asked John.

"He is gone forever and no; you are perfect the way you are. Don't ever be ashamed of it," assured his mother.

John closed his eyes and heard the ticking of a clock. He opened his eyes and found himself in a large maze, being chased by an eyeball. Then he saw a man whose face was covered. That man led John to a building, which looked like the post-apocalyptic world. The building was full of John's variations at different ages. He was afraid of being in the building, and so, he started to cry. Eventually, he woke up and found Hassan awake, next to him. Hassan comforted him back to sleep.

***

A new friend and some roommates

"Every new friend is a new adventure… the start of more memories." - Patrick Lindsay

Andrew wanted to message Ethan, but he wasn't sure if he should send it, even though he already had it written. He wanted to ask Ethan if he loves Hassan and what that meant for their relationship. Eventually, he got the courage to press send on the message. Ethan admitted his love for Hassan and they ended up taking a break but they decided to remain friends. Andrew opened Instagram to look at pictures of his ex. Looking at his pictures, he started talking to himself, "I loved you from the bottom of my heart but you broke it. You always made me cry. I lost my most precious years loving you. You took everything with you but left scars behind. Now, I am unable to open my heart to anyone." That night, a lot of people were in pain; John, Hassan, Cara, Ethan and Brian.

The good thing in Cara's life was Kyle; he was too sweet. The next morning, he prepared breakfast in bed for Cara. They had their whole weekend planned as well. Cara texted Hassan, telling him that she would be busy with Kyle over the weekend so that he could take care of the restaurant.   Hassan arrived at the restaurant but a young man stopped him at the door, asking if the restaurant was hiring. They were not interested in hiring, but Hassan wanted to help the man as he could see he was new to the city by the suitcase beside him. Upon further inquiry, Hassan found out that the guy's name was Pete and asked for his resume. Pete was fine doing any job as long as it would pay. Hassan asked Pete if he had anything for breakfast and he said no. Hassan stopped the interview so that he could have breakfast first. Once he was done, they continued the interview and found out that Pete was a transman. He had also attended cooking school and had some certificates from there. Hassan was impressed by his qualifications and asked him to cook a meal with dessert on the side. While cooking, Hassan noticed all the details like the way he used his tools, his cleanliness and how he worked on the presentation of the food. Hassan was impressed and the food tasted even better. Thus, Pete was offered a job in the kitchen.

Hassan questioned about his living situation as it was evident that he had no place to stay. After finding out about his situation, Hassan offered him to stay in his basement until he found a place to rent. Hassan also offered him advance payment so that he could settle in well. Pete was ecstatic at the offers made by Hassan and accepted it all without any hesitation. Hassan took him to the house and showed him the room in the basement. He gave Pete keys to the house and told him about a separate entrance. "I can't thank you enough for all that you have done for me," said Pete.

Hassan told John about Pete and he was okay with it. John went to see his mother while Hassan went upstairs to complete Pete's employment process. Pete had taken a shower and arranged all his stuff; he headed outside and found Hassan's atelier. He was mesmerized by Hassan's paintings and pictures, but he found one of them to be exceptional. It was a naked man sitting on the floor, his elbows between his hairy legs. His hands were on his head, covering the face except for one eye. Pete could feel that the man was either sad and ashamed, curious or sinister. As Pete delved into the picture, his tears started to flow like a waterfall. Hassan had come in time to comfort Pete by giving him a hug. Pete calmed down and asked Hassan about the man in the picture. Hassan told Pete that it was a picture of himself. Hassan asked why the picture meant so much to him; Pete explained that he could fully relate to the man in the picture since his father never accepted him for who he was. His mother was the only person who did but she had passed away.

"I'm so sorry but you don't have to be ashamed. You are a charming person. Anyone will be lucky to have you," said Hassan.

"My ex-boyfriend would disagree with you," replied Pete.

"He doesn't know what he's missing on. I have something that will make you feel good about yourself," said Hassan.

Hassan offered Pete to be a model for his pictures. Pete was very shy but he reluctantly said yes to the offer. He looked at Hassan's camera and remembered his mother's words, "You are my pride. You are my everything." It was a wonderful moment as Pete realized that he doesn't need to be ashamed of anything. He started posing like a supermodel. Hassan was smiling behind the camera. Pete decided to lose some of his clothes. Eventually, he was naked in front of the camera. Hassan took some shots from different angles. After a while, Pete took a long sigh and said, "Thank you." "Welcome to the family," replied Hassan.

Hassan offered him to live rent-free for one month and that he could move there as long as he pays the rent the next month. Hassan's actions left Pete thinking, "Who is this guy? Why is he helping me? I thought no one cared, but I was wrong. Thank you, Hassan. I don't know how I will be able to repay your kindness."

Hassan worked on the pictures using photoshop. He picked one of the pictures, printed it and gave it to Pete. Pete was amazed at seeing his picture; he couldn't believe how beautifully Hassan had captured him. Meanwhile, Ethan asked Hassan if he was interested in dinner, so Hassan asked him to come over to his place. Ethan asked if he could bring a friend along and invited Andrew. Hassan also invited Cara and Kyle. Ethan was happy that he didn't have to spend the night alone and Andrew was glad since he wouldn't be staring at pictures of his ex on Instagram.

Hassan asked Pete to cook dinner for the night and he also asked him to prepare desserts for all the customers of the restaurant; this would allow feedback on his work. They got all the groceries and went home to prepare for dinner. Hassan sat on the couch while Pete started working. Hassan opened up YouTube and played the first song that appeared, it was a Turkish song.

"I like this song. Come here and dance with me," said Hassan.

"I am not good at dancing," replied Pete.

"I don't know how to dance but who cares. Come on, don't make me drag you here," said Hassan.

They started dancing and Pete really started to enjoy himself. Hassan stopped dancing as tears started to fall from his eyes, Pete reminded him of his little brother. Pete looked at him and could feel the pain and sorrow. Losing someone you love is one of the worst things a person can experience in life. Getting through the loss takes a lot of time. Hassan moved towards Pete and held him between his arms.

"I promise, I will keep you safe," said Hassan as he kissed him on the forehead.

Pete was not sure what was going on and had a confused look on his face, and so, Hassan said, "Sorry if I scared you. You remind me of my little brother. I miss him so much." Pete asked about Hassan's brother.

"He left this world for a better place," said Hassan.

Pete apologized for asking and continued making dinner. As the time for dinner came close, all the guests had started to arrive. Hassan introduced everyone to Pete. Ethan's landlord wanted to do some renovations, which forced Ethan to look for a new place and so, he asked Hassan if there was any room for him to move into. Hassan asked Cara and John, they agreed and it was decided that Ethan would be moving in as well.

"I am the only one who doesn't live here," said Andrew.

"Kyle also doesn't live here," said John.

Kyle came and asked what they were talking about since he heard his name.

"We are talking about who is living here and who isn't," said Hassan.

"My girlfriend lives here, so technically, I do too," said Kyle.

Everyone laughed and welcomed Pete to their family. "Pete, how old are you?" asked Ethan. Pete told him that he was 22 years old.

"Oh, you are the youngest among us. John and I are 38 years old. Hassan is 34, Andrew is 33 and Kyle is 31 years old," said Ethan, then he whispered, "Don't ask Cara!"

"Hassan is 34? I thought he was in his twenties," said Pete.

"Oh, thank you. That is very sweet of you," said Hassan.

"Do you know what else is sweet?" John asked Hassan. Hassan nodded and so, John replied, "Your soft lips."

Hassan started blushing and asked John to stop. Cara told Pete that John always flirts like this with Hassan. Hassan took Ethan downstairs to show him his room. Ethan loved it and took the keys from Hassan.

"How much would the rent be?" asked Ethan

"We can talk about that later," said Hassan.

"I wanted to ask about your new roommate, do you know if he's single? Does he prefer men or women?" asked Ethan.

"You'll have to ask him that yourself. Are you interested in him?" asked Hassan.

"Yes, he seems different from other guys that I have been with," replied Ethan.

They heard that dinner was ready and went upstairs. During dinner, Brian was standing outside looking at everyone through the window. Everyone was busy eating, talking and laughing. Brian looked at Hassan and John, and his body started heating up. He closed his eyes and imagined himself instead of John. He started feeling himself thinking about being close to Hassan. "You belong to me," he whispered as he walked away from the window.

Brian entered his place and grabbed Hassan's painting that he had stolen. He held the painting close to himself and started talking to it, "You're not like other paintings. You tell my story without using any words. A story of deep pain and sorrow. Every time I look at you, you capture my heart. The pain in you is different. You were drawn when Hassan felt sorrow. Your pain isn't strange to me. I wish you could hear all that you mean to me, Hassan." He put the painting away and got a picture of Hassan, he put it close to his chest as he laid down and closed his eyes.

He opened his eyes and saw Hassan smiling. Brian pulled Hassan into bed and kissed him. "Come on. I want to show you something," Hassan said. Brian followed Hassan as he showed him some flowers in the yard. Hassan picked a flower and gave it to Brian. The moment Brian touched the flower, it turned to stone. He was confused. Hassan took the flower from his hand and the flower turned normal again. Hassan looked at Brian and pointed to a mirror. Brian walked to the mirror and stood in front of it. He looked at the mirror and saw John. He moved his right hand; John did the same. Brian freaked out. He ran but he couldn't find an exit. He found a wall and hit his head, losing consciousness. Brian opened his eyes and saw Hassan smiling. He pulled him into the bed but the bed turned to water. He was able to swim but saw Hassan drowning with a chain on his neck, hands and legs. He tried to help him but Hassan continued to drown. Finally, he woke up and then continued staring at the picture.

After dinner, Ethan followed Pete into the kitchen. Ethan complimented on Pete's cooking skills.

"Can I ask you a personal question?" asked Ethan.

"Sure, go ahead," Pete replied.

"Okay, this might be weird but are you in a relationship?" asked Ethan.

"No, I was in one, but it ended last month. I am not seeing anybody right now. I need time to put things in perspective," said Pete.

John screamed from the other room, "Ethan and Pete, come on, we're starting a movie."

They walked to the other room to join the others.

Ethan asked, "What are we watching?"

John replied, "The shape of water."

"It's about a unique relationship with an amphibious creature, right?" asked Cara.

John replied, "You are right."

"All of us need some of that," said Kyle.

John replied, "Speak for yourself, I need Hassan in my arms, that's all."

They played the movie and after 40 minutes, Hassan had fallen asleep. John took him upstairs and tucked him in. After the movie finished, Ethan wanted to talk privately with Cara about Hassan.

"You can talk in front of everyone, we are all involved in Hassan's life," explained Cara.

"Okay, then tell me what happened the day Hassan ran away," said Ethan.

Cara proceeded to tell everyone what happened that day in detail. The room was silent as she spoke. Most of them couldn't believe what they heard.

"That's very painful to hear, why would John do that?" asked Pete.

"That explains the crying. Hassan held the box in his hands like a little kid," said Andrew.

"I hate to say this, but I think John is trying to cover something up, it seems like he is spinning the truth," explained Ethan.

"What do you mean?" asked Cara.

"I have known him for years. He does things like this when he is hiding something big, he hurt Hassan so much that day, something he can never take back," explained Ethan.

"Why would you say that?" asked Cara.

"Hassan wanted to die that day," explained Ethan.

"That can't be. Hassan is full of life," said Cara in disbelief.

"Only John can do that. He gets blinded by his emotions and doesn't look at the damage that he causes," explained Ethan.

"I am so worried, what can we do?" asked Cara.

"Nothing," replied Ethan.

***

Chapter 6: Chaos

Friends like family

"Family means no one gets left behind or forgotten." – David Ogden Stiers

A few days later, Hassan and John were in their room cuddling as they had slept in each other's arms. John woke up first and saw Hassan sleeping peacefully. John kissed him and a smile appeared on Hassan's face as he opened his eyes.

"Good morning, shorty," said John as he put his hand on Hassan's chin.

"Hey, you," Hassan replied. "I always feel special when you wake me up with a kiss."

"I want to kiss you every day for the rest of our lives. We have the whole day to ourselves, by the way. What do you want to do?" asked John.

"I'm all yours then," said Hassan.

"I know a few things that we can do," said John as he pulled Hassan close.

"What will that be?" flirted Hassan with a coy smile on his face.

"You're about to find out," replied John.

John started kissing Hassan passionately on his lips and bite his lower lip. John took off Hassan's shirt and moved down to his shoulders madly kissing him. John bit him on his chest and moved down to Hassan's belly as he put Hassan's leg on his shoulder. John could hear Hassan moan and felt him quiver with passion. John reached up to Hassan so that they were face to face. John placed his hands around Hassan's neck and began to grasp it a little roughly. After a few seconds, John began to strangle Hassan until he could barely breathe. Hassan tried to free himself but was unable to as John's grip was strong. John tried to control himself; he tried hard but it seemed that he was not in his senses.

John closed his eyes and yelled, "I don't want to do this. Stop me."

"What are you waiting for? Finish him," John heard his father shouting at him.

John opened his eyes and saw his father sitting in front of him on a chair. He had a cigar in one hand and a drink in the other. John could sense his father's anger and felt terrified.

"This can't be real. You are not real..." fumbled John as he looked at Hassan, who had stopped moving.

"Please wake up. I'm so sorry," said John as he held Hassan in his arms.

"Why would you apologize? You have now become your true self. You are a monster," said John's father aggressively.

John started to scream repeatedly, "I'm not a monster! I'm not, I can't be."

Hassan woke John up at that very moment. "You were having a nightmare, wake up,"

"I'm not a monster," cried John as he placed his head on Hassan's chest.

"I'm right here. You are not a monster, John. It was just a dream," Hassan comforted him as he wrapped his arms around him and kissed him on the forehead. Hassan sang a song to John until he had fallen asleep and waited for the night to be over.

In the morning, Hassan woke up early, went to the kitchen and prepared breakfast. John was sound asleep when Hassan entered the room to wake him up. Hassan kissed him and asked him to come to the kitchen as he went downstairs to wake up Pete. John let out a stretch and a yawn, got up from the bed and washed his face. John looked at himself in the mirror as he observed his tried eyes. He shook his head and went to the kitchen.

"Good morning, slept well?" asked Pete cheerfully as John came downstairs and took a seat.

"I didn't sleep well. I feel tired as I had a bad dream," mumbled John as he took a bite of the toast.

Hassan looked at John with a concerned expression and said, "You will feel better after you take a nap in the afternoon."

"I will do that," said John as he gulped down his coffee.

"I was just getting ready and collecting my things. Do I have work today?" Pete tried to change the topic.

"I emailed you earlier. You are off today," replied Hassan.

"Come close to me. I want to feel your lips," said John as he pulled Hassan closer, put jam on his lips and gave him a long passionate kiss.

Pete was observing their love, reminisced his first kiss and thought to himself, "They look perfect together no doubt, but why does John have this obsession with Hassan, I can never understand."

As breakfast came to an end, John excused himself to work on his laptop to look at a few documents. Hassan and Pete began to clean up and started cleaning the dishes.

"You were looking at us when we were kissing, but you took a long pause. Why is that?" inquired Hassan as he put a plate back in the cupboard.

"I was thinking about my first kiss," sighed Pete.

"Why do I feel that there is a story that you are hiding?" asked Hassan as he stopped what he was doing and looked Pete straight in the eye.

With tears in his eyes, Pete said, "Before my transition, I had my first kiss with a guy. When my father found out about me, he sexually assaulted me to straighten me up."

"Don't cry, Pete. You can share with me. I won't judge you and will listen to you. I want you to release all the pain from your past and move on," said Hassan as he held Pete in his arms.

Pete started telling his story from the time he had his first sexual experience, all the hurdles he faced being Trans and how he lost his mom. Pete had finally uttered the suppressed feelings to someone other than his mom. Hassan couldn't believe a father could be this cruel to his son.

"I hate my father. I hate him so much," said Pete as he buried his face in Hassan's chest.

"When you hate someone, they are engraved on your heart. If you want to move on, you have to forget the person. Release yourself from the attachment and just let the past go. Forgive them, and only then you will feel unconditional love and oneness with everyone," said Hassan.

Pete thanked Hassan and wiped his tears.

"Can I ask you something, if you don't mind?" asked Pete. "Are you helping me because I remind you of your little brother?"

Hassan felt a sting in his heart, but he put on a brave smile. "You don't remind me of my little brother. I just saw him through you when you were dancing. I just like to help people and can't see them suffer."

"That's really kind of you," said Pete.

"What do you think about being part of my family?" asked Hassan.

"You don't even know me. Am I really worth it?" inquired Pete.

"You seem like a person who is worth knowing," smiled Hassan.

"That's so sweet. Yes, I will be a part of your family. I would love to be," Pete said.

"Welcome to the family," said Hassan as he patted Pete on the back.

***

No harm in having fun

"You don't get to choose if you get hurt in this world… but you do have some say in who hurts you." – John Green

As John was using his laptop, a message popped up on his phone. "Are you free to have some fun?"

John read the message and responded that he was not in the mood. He began to overthink and went to the private folder of his phone's gallery, which had naked pictures of Hassan sent to him by an unknown number.

"I can't believe you did this to me. It all starts from here, I don't know why but I will never forgive you," John thought and picked up his phone. He quickly typed a message that he will meet at 3 pm, stood up and headed to the bathroom. John took a shower and imagined Hassan naked. There were strange men with no faces touching Hassan all over his body. John turned around, put his hand on the wall and began to cry. He got out of the shower, put his clothes on and went downstairs to head out. John saw Hassan and their eyes met, but John looked away.

"I'm going outside. It will take me a while so don't wait up. I may stay at my mom's for the night," said John as he went towards the exit.

"That's alright but is everything okay?" asked Hassan.

John looked down before he said, "Everything is fine."

John had left in a hurry and it hurt Hassan quite a lot as he felt that there was something wrong. Hassan headed to the atelier, grabbed a pen and a paper from his notebook and started writing,

"Silence still takes over my life.

Pain is still a source of memories.

And sadness is still my music.

I wish fate can understand my words to witness my suffering.

But I will smile even if I am bleeding in pain."

John's photos were hanging on the wall and as Hassan took one from the wall, he said, "I'm watching you disappear."

As Hassan hugged John's picture, Pete knocked on the door and entered the atelier. Pete saw that Hassan was upset and asked him why he was feeling down to which Hassan explained that John doesn't share what's in his heart and it seems that he doesn't deem Hassan as trustworthy. Pete consoled him that there is love between the two of them and it will keep them strong. Hassan responded that love doesn't solve relationship issues and only sharing, along with trust, transforms a relationship. Pete went quiet for a few seconds and told Hassan that he was jealous of him because despite everything, Hassan had been strong. Hassan responded with tears rolling down his cheeks that behind the rough exterior is a man with immeasurable grief. Pete took Hassan in his arms and gave him a long hug. Ethan entered when Pete was embracing Hassan in his arms.

"Am I interrupting?" Ethan asked.

"We were just talking," said Pete as he let go off Hassan and left the atelier.

"Is everything okay?" said Ethan as he kissed Hassan on the forehead.

"Everything is okay," replied Hassan weakly.

Ethan picked up Hassan's notebook and saw what he had written. "Tell me the reason for your sadness. Is there something wrong between you and John?"

"I feel like John and I are drifting apart. The space between me and him is growing, and I don't know why," confessed Hassan.

"Do you want me to talk to him?" said Ethan. "He will be upset but he needs to be considerate of your feelings. Remember, I will always be on your side."

"I know, but don't do that as it will overcomplicate the matter," said Hassan. "When are you moving in though?"

"I will be living here for a long time. I hope you don't mind," said Ethan with a smile.

"I don't mind at all," Hassan smiled back.

"I love the big bed in my room. Before I forget, here's the cheque for the rent for my first month here," said Ethan as he handed the cheque to Hassan.

"I'm happy you like it. Do let me know if you need help to put the stuff in your room," said Hassan.

"I sure will," said Ethan as he headed back to his room.

After Ethan left, Hassan once again drowned in his feelings. To feel in control, he put on his MP3 player, grabbed his paint brush and started painting. The song began to play,

"Lithium, don't want to lock me up inside

Lithium, don't want to forget how it feels without

Lithium, I want to stay in love with my sorrow

Oh, but God, I want to let it go."

The world felt strange to Hassan. He felt disconnected as he felt himself getting lost in the music. As he drew lines and made brush strokes, choosing different colors, he felt his unspoken words were splashed on his painting. He turned the words in the mind into a precious piece of art. The song continued,

"Come to bed, don't make me sleep alone

Couldn't hide the emptiness; you let it show

Never wanted it to be so cold

Just didn't drink enough to say you love me

I can't hold on to me

Wonder what's wrong with me

Lithium, don't want to lock me up inside

Lithium, don't want to forget how it feels without

Lithium, I want to stay in love with my sorrow."

Ethan and Pete heard the song, entered the atelier and saw Hassan painting. Hassan didn't notice them and both of them stood there for hours as they saw Hassan turn his pain into a painting. After finishing his work, Hassan dropped the brush and backed up so that he could see his complete work.

Ethan walked up from behind and asked, "What do you call this? It's so deep."

"The Burning Wings," said Hassan and sat on the floor. "I can't feel my feet."

"The pain has consumed you and you spent all your energy. You used your emotions to draw and now you're feeling tired. Let's take you to your room," said Ethan.

Ethan took Hassan to his room and put him in his bed. He put a blanket over him and kissed him on his forehead. "Sleep well," said Ethan as he left the room.

"I'm worried about him. He is in immeasurable pain," said Pete.

"He will be alright," said Ethan.

"How can you say that? Did you even see him?" asked Pete.

"He is a strong man," said Ethan

Pete shrugged his shoulders and said, "Hammering iron makes it soft."

"I see your point," said Ethan

***

The past is history

"Live out of your imagination, not your history," – Stephen Covey.

The next day, Hassan got ready for work, had breakfast with Pete and since John hadn't shown up, he left a voice message on John's phone, "Sweetie, I haven't heard from you. Is everything okay? You didn't call or text me. Talk to me whenever you are free. Love you."

After arriving at his office, Hassan kept leaving messages for John but there was no response. Hassan couldn't concentrate on work as he kept thinking about John the whole day. After work, Hassan took an Uber home as he wanted to reach his place soon because he felt a hole burning in his heart. Hassan entered his home, sat on the couch and put his head back. Tears began to form in his eyes and dropped one by one. As Hassan wiped his tears, John entered through the door.

"Where have you been? I was worried," said Hassan. "I left you so many messages."

"You don't need to babysit me. I told you I will be at my mom's," said John.

"Is it something that I did?" asked Hassan as he tried to hold back his tears.

"Stop being so dramatic. I'm tired Hassan. Let me change and rest," said John and went up to his room.

Hassan waited for John to come down but he never did. Hassan then left for Lake Ontario Park. As he walked towards his destination, Hassan felt darkness looming over him and he couldn't control his tears. His friend sat next to him and asked Hassan whether John had hurt him. Hassan felt uneasy and responded that John has changed. Hassan's friend warned him that John was not good for him as he doesn't trust Hassan at all. Hassan explained that he felt lost and defeated to which his friend said that you are not lost, you just need to let go. Hassan just wanted to be left alone and asked his friend to free him. Hassan's friend explained that he will once Hassan himself is ready to let go.

At night, Hassan tried to reconcile with John by putting his arm on his shoulder, but John shrugged it off and put a cover on himself. The next morning, John woke up early and took a shower. Hassan didn't join him and waited for John to be done before he could use the shower.

At the breakfast table, Hassan kept looking at John but he didn't utter a single word. Cara and Pete were talking to Hassan but he couldn't hear anything as he was focused on John. As breakfast ended, John stood up to go to his car. When Hassan asked John to wait for him, John said to take the subway.

Hassan felt shattered. All the while, Cara tried to talk to Hassan and distract him, but Hassan took his bag and went outside to take a bus to the subway. As Hassan sat on the bus, he could see John and himself sitting together laughing and making jokes. John would smile at Hassan and kiss him from time to time. As Hassan got off the bus and headed for the escalator that took him to the subway, he again saw John and himself laughing and kissing each other. After every few moments, Hassan would see John and him holding hands, dancing, laughing and kissing.

As Hassan reached the station, he entered the subway, took a seat and closed his eyes. He saw himself on edge and was about to fall down when a guy, Devon, asked him something due to which he opened his eyes. Both of them talked for a while during the subway ride then went their separate ways.

After returning from work, Hassan was too overwhelmed and couldn't keep his emotions stuffed anymore. He let it all out and began to cry on the floor. John saw him, yet he said nothing instead and put a note on the refrigerator. "I will be at my mom's for the week."

After John left, Ethan entered a few moments later with a box full of his stuff to move into Hassan's place. Ethan saw in horror as Hassan wept on the floor. Ethan dropped the boxes and ran towards Hassan to comfort him. Ethan felt hurt and upset as he saw Hassan suffering. He took Hassan to his room and tucked him in his bed.

Ethan went downstairs and told Cara and Pete that Hassan wasn't his usual self and his soul was gone, which was painful to watch. Pete told Ethan that John had been rude and distant towards Hassan in the morning. Cara decided to head upstairs and talk to Hassan and make him feel better by setting up dinner. Hassan protested that he wasn't in the mood, but Cara explained to him that if someone doesn't treat us well, it is because they have a dysfunctional life themselves and Hassan doesn't have to suffer silently.

After dinner, they all decided to dance except for Hassan. Pete started moving funnily, which made Hassan laugh and he eventually joined everyone after a while. After cleaning up the kitchen with everyone and putting the leftover food in the fridge, Hassan saw the note left by John. He read it but just left it there.

After Cara and Hassan went to sleep, Pete went to take a shower. Ethan was putting his stuff in order and when he went to the bathroom, he saw Pete naked. Ethan apologized and left, but the image of Pete without clothes was stuck in his mind. Ethan went to Pete's room and gave him a kiss and Pete kissed him back. They made love all night and later slept together in each other's arms.

Hassan was in his room but couldn't sleep, so he typed a message to John but didn't send it. After closing his eyes, Hassan dreamt that he was on a cliff wearing all white with short sleeves and long colored strips wrapped around his arm. Hassan was on the edge and about to jump when someone stopped him by holding his hands. The person told Hassan that he needs to hold on and that they will meet in the future. The person told Hassan that he would be a great father and fix a broken family.

In the morning, Hassan was having breakfast when Ethan and Pete entered the kitchen.

"You both had sex last night," said Hassan.

Ethan was drinking coffee and spit it out of his nose to deny it. Hassan said that both of them were glowing and have the same smell, so there was no need to hide the truth. Hassan asked Ethan if it was a fling, to which Pete blushed and said that it was nice and refreshing.

Later, Ethan and Pete mutually decided to give their feelings a shot and go on a proper date. Meanwhile, Hassan met Devon at the subway station again. They talked for a while after which

Devon offered to walk with Hassan to his workplace. When they reached Hassan's office, Devon moved closer to Hassan to kiss him. John, who had come to meet Hassan, saw this and left, without even waiting for Hassan's reaction. Hassan, on the other hand, stopped Devon and made his intentions clear that he already had a boyfriend. Devon apologized and told Hassan that he will see him later.

As Hassan entered his office's building and took the elevator, Daniel was already inside and mocked Hassan for kissing other guys while being in a relationship. Hassan asked him to mind his own business and not to be judgmental. Hassan, without waiting for his response, exited the elevator. Daniel was fuming and went to his desk when Ethan asked both Hassan and Daniel to come to the conference room for a meeting. Daniel had to give a presentation, but it went terribly and the Head of Department asked Hassan to assist Daniel. Daniel took it personally and after the meeting, he lashed out at Hassan, who complained to HR and the Head of Department by sending them a detailed email.

Hassan took the rest of the day off. Since the last few months, it had been up and down with John, and it was really exhausting for him. Hassan arrived home and went to the atelier where he saw all his paintings and pictures destroyed. The colors were on the floor and his notebook was burned. Hassan was devastated and felt his face sweltering with pain. He heard a noise upstairs and forced himself to go and check it out, where he saw John breaking everything in their room. Hassan watched John acting on his anger, but stood still near the door as he destroyed everything.

John opened the closet and took out his small box and threw it out of the window into the yard. This made Hassan scream and he quickly left the room to find his box, but John held him by his neck and pushed him up against the wall. Hassan tried to free himself but eventually stopped and smiled, which further aggravated John and he released Hassan.

"I wish you were dead. I wish you were dead. I wish you were dead," John screamed at Hassan's face and walked away.

As John was leaving, Cara entered the house and saw Hassan on the floor with red marks on his neck. Hassan went barefoot to the yard to look for his box. After he found it, he went up to his room. Cara immediately called Ethan and asked him to come over as there was an emergency. Cara went to Hassan, who asked for a moment with his family's belongings that were in the box.

An hour had passed but there was no sign of Hassan. Ethan had driven his car like mad to reach Hassan's place, and when Ethan went up to Hassan's room, he found him bleeding as he had cut himself with a piece of glass.

***

Chapter 7: The Demon Under My Bed

Painful predicaments

"There are wounds that never show on the body that are deeper and more hurtful than anything that bleeds." — Laurell K. Hamilton

"The pain has gone," said Hassan weakly as Ethan ran to him and held him in his arms.

"Cara, come in here. Call 911 now!" yelled Ethan.

Cara entered the room with horror as she saw Ethan's shirt covered in blood.

"Why, Hassan?" asked Cara as Hassan smiled with his eyes full of tears.

Cara's tears began to drop on Hassan's face as she called 911.

"I'm sorry," said Hassan.

Ethan wrapped a cloth around Hassan's wrist to stop the bleeding as they waited for an ambulance. Meanwhile, Hassan saw his sister standing near the window, smiling.

"Finally, I will join you," said Hassan as he passed out.

Ethan and Cara were full of sadness and rage as they saw Hassan helplessly, but they didn't show it as there was a whole lot more that Hassan was feeling.

"Stay awake and live as you promised me!" yelled Cara as she heard the siren of the ambulance.

The paramedics arrived and took Hassan toward the ambulance. Ethan asked Cara to go with Hassan as he had something else to do.

"Don't do something stupid. Hassan only has you and me as his family," said Cara as she climbed the ambulance.

"Don't worry. I will see you at the hospital," said Ethan as he closed the ambulance's door.

Ethan sat in his car and drove straight to John's mother Marguerite's place. When he reached, Ethan started banging on the door wildly.

"I know you are in there! Open up, John," screamed Ethan as Marguerite opened the door.

"Where is that monster? I want to see him right now," said Ethan as he frantically looked around for John.

"You have no right to come here and speak about my son like this," exclaimed Marguerite as John entered the room.

"Whose blood is that on your shirt?" asked John with worry.

"Do you really want to know?" said Ethan.

"Answer me!" yelled John.

Ethan explained that Hassan had cut his wrists, was taken by the paramedics and wasn't sure whether he would make it.

"He had signed the DNR last month. I killed him," said John palely.

Ethan's eyes began to fill with tears and as he was about to leave, John asked which hospital Hassan was taken to.

"You have done enough. I could kill you right now. Hassan doesn't deserve this. He has only ever loved you. Stay away from him. You are a living nightmare," said Ethan.

After Ethan left, Marguerite remarked that John had indeed become a monster just like his father, to which John felt helpless.

Meanwhile, Ethan reached the hospital in his car and went to the Emergency Room to look for Cara.

"Is he alright? Did you know Hassan signed a DNR?" asked Ethan.

"The doctors are still with him. Yes, I know. He asked me to mail the DNR form, but I never did," said Cara.

"I'm glad, thank you," said Ethan as he hugged Cara.

Ethan told Cara that he went to John to tell him off and stay away from Hassan. Cara responded that she would never let John near Hassan as he had done enough damage. Ethan explained to her that Hassan will forgive John anyway and that both Ethan and Cara had to support Hassan on whatever he decides. Cara protested that Ethan and herself have to look out for Hassan, to which Ethan agreed but argued that it was all in Hassan's hands.

As Ethan and Cara sat in the waiting room, the doctors arrived and informed that Hassan had a cardiac arrest but was out of danger and the next 24 hours were crucial, therefore, Hassan will have to be kept under observation. Ethan and Cara took a sigh of relief and hugged each other.

***

Reminisce and let go

"He was both everything I could ever want…

And nothing I could ever have…"

— Ranata Suzuki

John was driving his car and frantically calling Ethan and Cara, but they weren't responding. Since he couldn't be with Hassan at the moment, he went back to the house he shared with him. As he entered through the garage, John saw Hassan's fingers on the wall. John went downstairs and saw the paintings and photos that he had destroyed. He had once lived happily in this space and now everything was ruined. As John cried, he noticed Hassan's torn notebook lying on the floor. A page that was still intact said,

"You are here, but you are not with me. It feels like a dark cloud has been passed on in our relationship. Your hands dead in mine. I am with you, but only your shadow is here. I am watching you disappear, leaving me alone. Loneliness is nothing I want. Loneliness is shut-off walls, locked windows and doors. Without you, I am fading away. Without you, I cannot stay. The black cloud took everything from me, took everything I had. Now the clouds cry for the both of us."

John hugged the notebook close to his heart as he noticed a picture of him and Hassan smiling at the camera. John closed his eyes and saw Hassan smiling. He tried calling Ethan and Cara yet again, but there was no response.

***

Born to live

"The two most important days in your life are the day you are born and the day you find out why."

— Mark Twain

Brian called Cara to check on her as he did from time to time. Cara told him that Hassan was in bad condition. He insisted that he should visit, to which Cara refused, but Brian pestered and Cara finally gave in and texted him the address of the hospital.

When Brian arrived and saw Hassan, he began to cry and kissed his forehead. Hassan was still unconscious, so Brian left after a while. After a few moments, Pete also reached and sat beside Hassan as Ethan and Cara prayed for him to get better.

Hassan, on the other hand, was sound asleep in his bed and saw a white foggy place in his dream. He was looking around when he saw his father standing faraway. Hassan went toward him, but he walked away. Hassan was startled and shook his head when he heard his sister's voice.

"He doesn't want to talk to you," Hassan's sister said.

"I'm glad to see you," said Hassan.

"I'm not glad to see you at all. You must return," said Hassan's sister.

"I have to join you. You are my family," said Hassan.

"I don't want to do this but no one wants to talk to you here. Father is angry at you," said Hassan's sister.

"I want to be with you," said Hassan.

"I don't think so. Look over there," said Hassan's sister as she pointed to a small dug hole holding water.

In the water, Hassan could see Cara, Ethan and Pete crying and praying for Hassan.

"You are not the kind of person who turns his back on his family," said Hassan's sister.

"You are my family," said Hassan.

"They are your family now and you must go back to them. You belong with them. Go back, take care of them. We will meet someday, but not today, not this way," said Hassan's sister as she walked away.

Hassan woke up and felt Pete holding his hand. Hassan saw Ethan by the window crying as Cara sat on a chair praying.

"I'm sorry," said Hassan weakly.

"Don't ever do that again," said Pete as he kissed Hassan's hand.

Hassan asked about the whereabouts of John, to which Cara got upset and asked that after all that John had done, why did he still want to see him. Hassan explained that he wanted to end his misery once and for all. After a few minutes, Ethan called John and told him to come to the hospital.

As John entered, Ethan was punching the wall and Cara looked at him as if she wanted to slap John, but Ethan restrained her. Pete, however, punched John with all his might.

"I deserve it," said John as he began to cry.

John went towards Hassan and began to apologize but Hassan interrupted him.

"I have already forgiven you, but can you forgive yourself for what you have done?" asked Hassan.

John was taken aback and responded that he can't live without Hassan.

"My wounds are still fresh and I can't take it anymore. This is your last chance. But before that, do you have anything you want to say?" asked Hassan.

John shook his head.

"Remember, I gave you a chance to explain yourself. If I ever find out that you lied to me, we will be done," said Hassan.

"I will never hurt you again," said John.

"I want you to prove it through your actions as I have already heard this before. Go home, clean up the mess and we will live in separate rooms until I can trust you again," said Hassan.

As John left, Hassan asked Ethan and Cara to go home while Pete would stay with him in the hospital. After Cara and Ethan left, Pete put his head on Hassan's shoulder and laid beside him in the bed.

"What happened between you and Ethan?" asked Hassan to release the tension.

As Pete went on about Ethan and his encounter with him, Hassan listened silently and smiled.

"Will you let John back into your life?" asked Pete after he finished his story.

"He hasn't tasted my punishment and if he ever hurts me again, I will leave him. In the end, I'm no angel," said Hassan.

"Should I be scared?" asked Pete.

"You shouldn't be, but there are so many doors for John. Some of them opened with pain, some of them opened with anger, but most of them opened with love. I cannot hate him even when he hurts me. Every door I have for him has one key. He is the key to every door I have for him, and I will have more doors for him. Every door has a special part of me. I will always keep them in me," explained Hassan.

Pete responded that every guy he has dated treated him as a sex object and never as an equal. Hassan advised him that he should walk away from situations that cause harm and must set himself as a priority. Pete begins to cry and calls Hassan his brother. Hassan wipes his tears and holds Pete's hand and tells him that he will always find a brother and family in Hassan.

As Hassan and Pete were talking, Hassan received a text from John that said that he wanted to talk on the phone. Hassan agreed, and as John called, he gave him a warning that he had awakened a monster in Hassan and if John hurt him again, he would be severely punished. John got confused but Hassan continued that he will make him feel the pain that he was feeling right now.

John wanted to heal Hassan, but he was shattered into a million pieces and asked John if he could actually handle love and could it exist without pain.

John explained that he saw Hassan kiss a guy and got angry, due to which he destroyed everything. Hassan told John that he was blinded by jealousy and if he had stayed another moment, he would know that it was all a misunderstanding. After ending the call, Hassan began to cry and Pete wrapped him in his arms.

After a few hours, Kyle visited Hassan and when he saw that he was asleep, Kyle began to relive the memory when he had met Hassan for the first time.

Kyle was going to propose to his girlfriend but he found her sleeping with his best friend on his bed. Kyle was crying and walked aimlessly on the street when he decided to sit on a walkway. No one noticed him until Hassan came and sat next to him.

"If you want to cry, cry like a winner. I don't know what happened to you, but crying here like this won't help you," said Hassan.

"Leave me alone," Kyle said.

"Stand up. We can go to a park nearby," said Hassan.

Kyle wanted to stay, but Hassan pestered him and after a few minutes, both of them walked towards the park and sat on a bench.

"You are upset because of a girl," said Hassan.

"How did you know?" asked Kyle.

"You're holding an expensive ring, at least it looks expensive," said Hassan.

"I found my girlfriend in bed with my best friend," said Kyle.

"First of all, you have to return the ring and make your girlfriend move by collecting her stuff, putting it outside your door and changing the locks," said Hassan.

"But where will she go?" asked Kyle.

"Look at yourself! You are worried about a girl who cheated on you. She can live with your ex-best friend. Who cares?" said Hassan.

Kyle and Hassan went to the jewelry store to return the ring and after that, Hassan asked Kyle to smile and said that it would make him unstoppable as anyone who smiles with a broken heart can do anything. Kyle thanked Hassan profusely and said that Hassan made his day and he felt much better.

The next morning, Hassan woke up and saw Pete sleeping beside him. After a while, Hassan woke Pete up and thanked him for being there for him, to which Pete said that they were family. After a while, Brian came to visit and Pete left for the cafeteria. Brian handed a bluish-green crystal called the Amazonite to Hassan and told him to rub it between his finger and thumb for five minutes whenever he felt overwhelmed. Hassan thanked him and Brian gave him a kiss on the forehead. After Brian left the room, he thought for a moment about how he took advantage of Hassan but wasn't feeling guilty.

After Brian left, Kyle came to visit and asked Hassan whether he was doing better, to which Hassan responded that he was feeling weak. Kyle said that this should be John's last chance to which Hassan agreed. After a while, Kyle left and Pete came back from the cafeteria asking Hassan to get better so that they can have a big hearty breakfast at home.

Later in the day, Andrew visited and talked to Hassan about his health. Hassan assured him that he was doing alright and Andrew suggested that he should move in as he wanted to be around people who were caring and compassionate because his ex, Patrick, had done a number on him. Hassan told Andrew to move in anytime he wants.

John visited Hassan in the hospital with flowers and began talking when Hassan interrupted him.

"People in relationships are expected to be the closest allies, most incredible love and support sources. I love John, who is the love of my life. The kindest, most fantastic person in the world. You are not him," said Hassan.

"I'm right here. I'm with you," said John as he wiped his tears.

"If you are, then why do I have this?" said Hassan as he pointed to his wrist.

"I understand you are upset," said John.

"Upset? I'm furious. You don't get to slaughter me with a blind blade and return to my life like nothing happened," said Hassan.

"You can do whatever you want with me, but please don't leave me," pleaded John.

"I want you to feel what I felt. I want you to feel the pain you've caused and see my wrath, so you will never do this again," said Hassan.

John felt guilty and helpless, and Hassan asked him to sleep in the guest room for a while before things got better. Hassan was getting discharged tomorrow and John asked him if he wanted help getting home, to which Hassan responded that he will see him once he gets home.

John was about to leave when Hassan whispered a slight 'I love you.' John turned around and gave Hassan a long comforting hug with a kiss on the forehead.

"Is there anything you want to say before we move on?" asked Hassan

"Nothing at all," John responded.

"Remember, this is your last chance. If I find out that you hid something or lied, that means we are done," said Hassan.

"I understand," said John.

John left the hospital thinking that he had lost Hassan forever as it was only a matter of time before Hassan left him. The fear of being alone and the guilt of hurting Hassan was eating John from the inside. John didn't want the relationship to be over, but it was now out of his hands.

Hassan was alone in the hospital room with tears in his eyes.

"I am sorry for being hard on you, but I need our relationship to rise from the dead. Please don't hide anything. I want to be with you," Hassan said aloud.

People tend to forget about the relationship they have with themselves and completely prioritize the other person, which isn't fair. There are things that are in a person's control and then there are somethings that are not. There are so many unexpected internal shifts that can happen when a relationship is being repaired. It is only a matter of time until the picture gets clearer and things sort themselves out.

It was time for Hassan to leave the hospital and return home. Andrew had come to pick him up along with Pete, who was quite happy to go home as he hated hospitals.

"I am happy that you are leaving the hospital," said Pete.

"I know. I can sense it," said Hassan with a smile.

"What's the first thing you want me to cook for you when we get home?" asked Pete.

"Do you think of food all the time?" asked Andrew coyly.

"Not all the time. I'm just being polite," said Pete.

"I was only joking with you," said Andrew.

"Really!" said Pete.

"Don't mind him. I would love to eat anything you want to cook," said Hassan and smiled.

"Okay, but you have to eat some soup," said Pete.

"Deal," said Hassan

Pete and Hassan got in Andrew's car and he drove home. On the way, Pete asked Hassan when they will go on a city tour, to which he responded with a 'soon'.

Once they reached home, Hassan got out of the car, but instead of going inside, he stood in front of the house, looked at it silently, and everything that had happened on the day he slit his wrists came rushing back. It was a painful time for everyone involved and Hassan never wanted such a situation to ever happen again.

***

Chapter 8: Back Home

A place of love

"Home isn't where you're from, it's where you find light when all grows dark."

— Pierce Brown

Hassan truly understood the meaning of this quote for the first time as he saw all of his friends, who had now become his family, standing at the door to welcome him back from the hospital. Cara had opened the door before they even had a chance to ring the doorbell as she heard Andrew's car in the driveway.

"Welcome home," everyone said in unison, to which Hassan smiled and gave each of them a hug as all of them had made an effort to throw him a welcome party. After a while, John showed up, which made Ethan angry and jump up from his seat.

"What is this monster doing here?" Ethan asked Hassan.

Hassan signaled Ethan to sit down and went towards John, who was crying that he isn't a monster.

"You are not a monster, John. You just made a few mistakes, that's all," said Hassan.

John hugged Hassan, but Hassan stood still and didn't wrap his arms around him like he used to in the past.

"I almost forgot that my mother is here as well. Can she come in?" asked John.

"Marguerite is here? Of course, she can," said Hassan as he nodded his head.

John went outside to call his mother and both of them entered the house together. Marguerite inquired Hassan about his health, to which he responded that he was doing well. Hassan asked her to stay the night, but Marguerite felt that she wasn't welcomed. Hassan asked everyone whether they were comfortable if Marguerite stayed the night. As nobody wanted to upset Hassan, they all agreed. Marguerite began to cry and Hassan wiped her tears.

"You are always welcome here. We all love you," said Hassan to Marguerite as he hugged her.

Pete walked up to Marguerite and began to ask her about her cooking.

"We haven't met before. I'm Pete and I heard a lot about your cooking. Hassan loves the garlic honey chicken you make. Can you please share the recipe with me?" asked Pete.

"I would love to. Let's go to the kitchen and we can make it now," said Marguerite.

"That would be great. Thank you!" responded Pete.

Marguerite and Pete went to the kitchen to prepare dinner as John and Hassan began to converse.

"If my mom will stay in the guest house, then where will I stay?" asked John,

"You can stay in our room. Just for tonight," said Hassan.

"I thought we were over," said John.

"We didn't say that. Did we?" asked Hassan.

"We didn't," said John.

"I'm glad that's cleared out between the both of us," said Hassan.

Hassan excused himself as he had to show Andrew the room that he would stay in at Hassan's house. Hassan collected the keys to the room from the kitchen drawer and both of them went downstairs. Hassan showed Andrew his room which took him by surprise as it was fully-furnished and decorated quite nicely.

"I won't have to bring my furniture from the condo. You have amicable taste, by the way. I love it," said Andrew.

Hassan explained to Andrew that he had an attached private washroom with his room while Ethan and Pete would be sharing a bathroom. Andrew asked about the rent and Hassan responded that it would be decided when Andrew moves in. Andrew was grateful and thanked Hassan for his kindness.

Hassan walked to the atelier, saw everything that was destroyed and felt pain rise in his chest.

"Did John do this?" asked Andrew.

"Yes, he did," said Hassan.

Andrew hugged Hassan and he began to cry silently.

"Let it out. It does hurt, but after a hard, loud, wailing, sorrowful cry, we feel better. We let the burden go. When we're empty, we feel lighter. I hate to admit it, but I got pretty skilled at crying silently over the years. Don't keep it in; let it out and heal," said Andrew.

Hassan held on to Andrew and began to cry loudly as he consoled him.

"This shirt is quite expensive and now it is ruined. I will deduct its cost from the rent," joked Andrew.

Hassan laughed out loud.

"It's all about the rent for you," said Hassan as a smile came across his face.

"Wash your face and then we will go upstairs," said Andrew.

Hassan quickly washed his face and went with Andrew upstairs to join the others. The song Hey Mama by David Guetta began to play and Cara asked Hassan to come and dance. Andrew asked John to go dance as well, to which John complied and joined Hassan. As the song ended, John asked if he could choose the next song. Meanwhile, Pete and Marguerite returned from the kitchen and joined everyone. John decided to play If It's All I Ever Do by Anders Johansson. The song blasted from the speakers.

"Ooh, my love

The first time that you spoke my name

It somehow sounded not the same

Was like I knew from that moment on

That this is what I'm living for

Fate had opened up the door

And who am I to say that heaven could be wrong."

John put his arms around Hassan and danced with him. Ethan took Pete's hands while Cara and Kyle began to dance together. Andrew asked Marguerite to dance with him, and she agreed.

"If it's all I ever do

I would give my heart to you

And I will do it faithfully

Until the end of time

When they carve my name in stone

At least I know they'll know

That in this life I've made mistakes

That I did one thing right

Cause I will spend forever loving you

If it's all I ever do, my love."

John looked at Hassan's face and sang with the song.

"Baby, when I look at you

Standing there so pure and true

I don't know what I did to deserve

The way you smile

The way we touch

The way you kiss, it means so much

I must be the luckiest man

In the whole wide world."

John meant every single word of this song. He wanted to show Hassan how much he loved him.

"And you'd do anything on earth for me

And it makes me love you more

Cause baby, you're the only love I need".

John got closer to Hassan and kissed him on the lips.

"I love you, shorty," said John.

Ethan kissed Pete and told him that they must go on a date as Ethan wanted to know him better. Whilst, Kyle held Cara close to his chest and told her he is falling for her, and she is the most incredible woman he has ever met. Cara held Kyle with her hands and held on to him. It seemed like a house full of love, warmth and energy and happiness surrounded them.

It was getting late and soon it was time to sleep. Cara and Kyle went together to her room. They were lying on the bed, looking at each other. Kyle touched Cara's hair with his hand, then got closer and kissed her.

"I want to look at you the whole night," said Kyle as Cara laid between Kyle's arms.

"I want to sleep in your arms," Cara said as Kyle held her in his arms and asked her to close her eyes.

In the guest room, Marguerite prayed to God before she went to sleep. She put her head on the pillow and got apprehensive about what will happen next with John and Hassan. She had this feeling that something terrible was yet to happen. Finally, she closed her eyes, gave in and fell asleep.

Andrew was in his room and excited about the new journey he was about to partake in Hassan's house. Meanwhile, Ethan entered the room that he and Pete shared in only his underwear.

"What's going on? Why are you walking around in your underwear?" asked Pete as he got up from the bed.

"I like to sleep in the nude," said Ethan.

Ethan undressed Pete slowly, so he enjoyed looking at every inch of Pete's body. Then he held him between his arms and put him in the bed as they felt each other skin. Pete felt Ethan's warm hands moving around his body. Pete liked what he was feeling. Ethan cuddled Pete and they fell asleep together.

***

Destiny calls

"It's far better to be unhappy alone than unhappy with someone." — Marilyn Monroe

In Hassan and John's room, both of them slept on the opposite side of the bed. John wanted to touch Hassan, but he was contemplating if he should as he knew Hassan wanted his space. Finally, John moved closer to Hassan and put his arm around his waist. He felt Hassan's heart beating faster and realized Hassan wasn't asleep. He moved Hassan on his back, moved on top of him and kissed him.

"I missed you so much," John said.

They started kissing passionately and it seemed their relationship was taking a new direction. John put his hand on Hassan's wrists and moved his hands on the top of his head.

"Stop, you are hurting me," yelled Hassan.

John realized Hassan's hands were hurting as they were wounded and apologized to him.

"Get off me," yelled Hassan and John moved aside as he looked at Hassan with tears in his eyes.

"I am so sorry. I didn't mean to," said John as Hassan got up and walked towards the door.

"Don't follow me," said Hassan.

"What is happening to me?" John said as Hassan left the room.

Hassan went downstairs to Andrew's room and knocked on the door. Andrew got up and opened the door to let Hassan in. Andrew asked Hassan if everything was alright and Hassan responded that his wrists were hurting him as John had grabbed them aggressively by accident.

Andrew offered to check his wound, changed his bandages and showed his concern about John being careless with Hassan. Andrew insisted that Hassan sleep in Andrew's room today and not take any risk by making excuses for John.

Upstairs, John was crying silently. He put his face in his pillow so that nobody could hear his screams. He felt the space between Hassan and him growing deeper. His communication tactics lately weren't helpful in expressing his feelings to Hassan. John was incredibly aware and sensitive to the gap between what he said and how it's interpreted. He felt the deep connection that he and Hassan had begun to vanish and fade away. He put the pillow aside as he was now exhausted from all the crying and fell asleep.

Andrew woke up in the middle of the night from a nightmare and noticed that Hassan was breathing heavily. Andrew checked his forehead and felt that Hassan had a fever. He put a wet towel on his forehead and gave him an extra blanket just in case.

Soon it was morning and John saw that Hassan hadn't returned to the room. John went downstairs and began to search for him but couldn't find him anywhere. John began to imagine Hassan naked with Ethan, Pete and Andrew and felt his head burn. Just then, Andrew came upstairs and told John that Hassan had a fever and was feeling better now. Andrew asked John to control his actions and told him that Hassan was madly in love with him but needed some space and time from John.

Andrew left and John went to his room to get Hassan, but he was not in a good mood.

"Good morning," said John with a smile.

"Good morning," responded Hassan coldly.

"I was about to take you to your bed," said John.

"Thanks," said Hassan.

Hassan got out of the bed, walked to the door and went upstairs. John followed him as Hassan went to his room to change.

"Do you need help?" asked John.

"I am going to take a shower first. And I don't need your help," said Hassan.

"I am sorry. I didn't mean to hurt you last night," said John.

"Are you done apologizing?" asked Hassan angrily.

"I don't know what to say," said John.

"I don't need an apology. Your apology won't change what you have done," said Hassan.

"I know my apology doesn't change the past, but I'll do better. I promise," said John.

Hassan told John that he needed to show him through his actions and not just words. After that, Hassan walked away and went downstairs, where he found Marguerite in the kitchen. She asked Hassan if he slept well, to which he responded that he had slept enough. Marguerite then asked Hassan to promise her that he will forgive John. Hassan said that he can't promise anything right now. Marguerite said that it might seem difficult to forgive someone that you are deeply in love with as she had this experience in the past with John's father.

Hassan assured Marguerite that John was nothing like his father and that he had made a few mistakes. Marguerite made Hassan promise that even if he and John broke up, Hassan would not give up on him and treat him like family. Hassan agreed and promised that he would be there for John no matter what.

A week had passed, yet Hassan wouldn't speak to John and even when he did, Hassan would say spiteful things to hurt him. John would cry every night and tried every day to mend his relationship with Hassan, yet he paid no heed. Everyone in the house asked Hassan to treat John better as he was getting depressed and low, but Hassan remained unmoved and said that he knew what he was doing.

It was a Saturday morning when John woke up early and decided to make everyone breakfast. He wanted to talk to Hassan and ask him to end his mistreatment towards him. John woke up Ethan, Pete, Andrew and Cara and asked them to join him in the kitchen. Everyone was afraid of the conversation that was about to unfold.

"Who prepared this?" asked Hassan.

"I did," said John.

Hassan left the room after taking out a knife from the cabinet to cut the fruits as tears flowed from John's eyes. Hassan hit his hand hard on the table as John saw him and began to yell.

"You told me that you had forgiven me then why are you treating me like this?" asked John loudly.

"Why are you yelling?" asked Hassan.

"Do you want to cut my hands with this knife so you will forgive me? Or at least treat me better?" asked John.

"Treat you better! I begged you to give me the box. What did you with it? I can still hear you saying, 'I wish you were dead.' Your hands were on my neck, and you kept saying it. Why don't you finish what you started?" said Hassan.

John began to shake.

"Finish what you started," yelled Hassan as he put the knife in John's hand and held it against his own chest.

John dropped the knife and fell on his knees. Hassan had become cruel and cold hearted and nobody could believe how Hassan was treating John but didn't say a word. They couldn't recognize this Hassan, who was standing in front of them. John tried to cut his wrists, but Hassan stopped him and whispered something slowly in his ear, which made John smile.

"I love you," said John as he stood up and Hassan hugged and kissed him.

Everyone was staring in shock at what had just happened. Hassan took John's hand and went to the table to have breakfast and asked everyone to join. Hassan ate the breakfast even though it didn't taste right. Everyone obliged as they didn't want to upset Hassan. After a while, John took a bite, said that the breakfast is terrible and asked everyone why they ate it.

Cara said that Hassan forced us, to which Hassan responded that John cooked this for me and I will eat it with a smile as you all will too. John said he was tired and Hassan asked him to go upstairs to rest, and that he will join him shortly. After John left, everyone asked Hassan about his changed behavior.

"You went too far, "said Cara.

"It was hard to watch," said Ethan.

"I couldn't recognize you. You were a different person. Be careful how you treat John as he is madly in love with you," said Andrew.

"What did you even whisper in his ear anyway? It made him stop," asked Ethan.

"I just reminded him of something," said Hassan.

Andrew said that he understood, but he would never forgive Hassan for making him eat the salty breakfast. Hassan said that he never forced anyone to eat, to which Cara said that they were afraid of him. Hassan excused himself and said that he must go upstairs to his boyfriend and everyone cheered him on and wished him luck. After reaching upstairs, Hassan's eyes met with John.

"Try to sleep," said Hassan.

"Where is my kiss?" asked John.

Hassan gave him a long passionate kiss, after which John closed his eyes and fell asleep. Hassan kissed him on the forehead.

"I'm sorry I was cruel to you," whispered Hassan in John's ear.

After a while, Hassan went to take a shower and John woke up upon hearing the sound of water. John quickly undressed and joined Hassan. He was standing under the water and his head was up. John came from behind and wrapped his arm around Hassan's chest. Hassan moved his right hand on John's head and moved the other hand on John's leg. John bit Hassan's neck and kissed him on his neck as John and Hassan faced each other. John grabbed Hassan's butt; Hassan did the same. John looked at Hassan's face and wrapped his arms around Hassan's waist and moved him up. Hassan wrapped his arm around John's neck and his legs around John's waist. John was able to hear Hassan breathing, and his heart was beating fast. John was breathing slowly, but his heart was beating like crazy.

Hassan gave John a long, deep and passionate kiss which made John lose control. They kissed for a few minutes as Hassan used his foot to turn off the shower tap. John understood that Hassan wanted to move to the bedroom. John dropped him on the floor, then Hassan grabbed a towel and dried John while did the same for Hassan. He then carried Hassan to the bedroom and threw him on the bed. John laid on the bed next to Hassan. He moved on top of John and started kissing. Every kiss was saying I missed you and my heart is beating for you. Every kiss was telling a love story.

John flipped Hassan on the bed, put his hand on Hassan's hands and tangled their fingers together. Hassan used his teeth on John's chin to drive him even crazier. John moved down to Hassan's neck, chest and nipples, waist, and to every inch of Hassan's body. John sat on Hassan and started moving up and down. Hassan grabbed the sheets with his hands, then looked at John and flipped him. He put one of John's legs on his shoulder and the other leg on John's chest. He kept moving while John was growling. John felt Hassan had finished, but he kept moving. Hassan stopped and got close to John's face and kissed him and moved down.

John was enjoying what Hassan was doing. Hassan moved on to John's body, and when John was inside Hassan, he grabbed his waist tight with both hands as he moved up and down and increased his speed. He looked at Hassan and saw the view he won't ever forget in his lifetime. Hassan put his hands on John's body. John started growling loudly until he finished. Hassan put his head on John's chest, and John moved his hands on Hassan's body. He felt Hassan's skin and realized that Hassan finished too at the same time together.

"That wasn't expected, but it was wonderful," said John as he wrapped his arms around Hassan's back.

Meanwhile, Ethan and Pete's relationship progresses as Ethan gives Pete a hand watch as a gift. Pete becomes emotional and begins to cry as Ethan hugs and consoles him. Pete then tells Ethan that he was the first person who treats him equally. Ethan says that the first time he saw Pete, he went crazy as he was amazing. Both of them make love and then fell asleep, holding each other close.

Chapter 9: Loss

Learn from Loss

"Mostly it is loss which teaches us about the worth of things." — Arthur Schopenhauer

Andrew brought his date to the house and as they entered, they ran into John. Andrew's date recognized John. He walked to him and said, "Hi, John. Do you live here?"

"Alex! What are you doing here?" John got shocked seeing Alex in his house.

"I am Andrew's date. This should be fun, if you know what I mean," Alex teased.

"Oh, I didn't realize you both had…," said Andrew.

John took Andrew aside and asked him to take his date out before Hassan comes down. Andrew talked to Alex and told him about the complexities of Hassan's and John's relationship. He also asked him not to mention anything in front of Hassan. Alex promised that he wouldn't.

They were all talking when they heard Hassan's ringtone as he was coming down the stairs. Andrew took this opportunity to grab John aside and asked, "Are you cheating on Hassan?"

"Not the time. I will tell you later, but please make him go," John pleaded.

"You better have a good explanation," said Andrew while making his way to Alex.

They went outside. Alex asked Andrew if he could leave, and Andrew walked him to the door. Andrew came back and said, "You both need to go out for a date."

"You can come with us," Hassan offered.

"To hold a candle!" Andrew said sarcastically.

"You can come with us; it is not a date," Hassan insisted.

"Can everyone come downstairs?" Andrew called out.

"Sure, we will take a city tour, then go to a good restaurant," John said, looking at Hassan.

"I want to go to Nie," Andrew exclaimed.

Hassan asked John to inform Kyle while he dialed 'big Mike.'

"Who is 'Big Mike'?" asked Andrew

"One of my best friends. He taught me how to be a successful manager," said Hassan.

"I assume there is a story behind it," Andrew questioned.

"You will like it," John interrupted.

"Let Pete and Ethan know," said Hassan.

John and Hassan were having a fun banter where everyone laughed. While laughing, Andrew said, "What a weird combination, one with a British accent. The other with a weird accent and voice. You both are great together. I hope I can find my weirdo."

"Someday, you will find him," John said, taking Hassan's hand in his.

"Chop chop, time to get dressed," Hassan stood up with John. Hassan put his hands around John's neck and kissed him, but suddenly he stopped. He put his face on John's chest and said, "Ask him to stop." John looked over at Andrew, and he saw him touching Hassan's ass.

"What are you doing?" said John.

"That is a quality ass. You are very lucky," Andrew winked.

"Do you want to keep your hands?" John said.

"Woah! Why is he hiding his face in your chest?" Andrew asked.

"He is embarrassed because of what you did," John responded, looking at Hassan.

"Oh, really!" Andrew said sarcastically.

Everyone went to their room to get dressed for the night. Andrew undressed and entered the shower. With every drop, he felt the past washing away. He had always been shy; he was the guy who never had any jokes. Now he was the guy who would tell jokes all the time. A thought crossed his mind and he laughed, "I think I have to thank Hassan's clots," Andrew said.

Pete, Ethan, and Andrew got dressed and went upstairs to find John and Hassan dancing.

"How could you dance without us. From the start, please," Andrew said as he entered the room.

Everyone laughed. John played the song again, and everyone started dancing. Andrew looked around and told himself, "This is what home looks like. A real home with people who care about me." Ethan was leaning in to kiss Pete, but Andrew stopped him and said, "Don't kiss him, or I will have to kiss all of you." Everyone looked at Andrew.

"What! I am kidding," Andrew moved back and smiled.

Everyone laughed and continued to dance.

After this small dance party, they went to the shopping center, they spent some time there and met Cara and Kyle. Once they were done with the shopping, the group divided and went to the restaurant in John and Andrew's car. When they arrived there, they met Mike and Marguerite, who were already there. They all sat on a big table that was prepared for them. Hassan raised his glass and said, "To us, to family." Everyone raised their glasses. They were all chatting with each other while waiting for their food to be served. Mike looked at Hassan and said, "I need you to sign some papers tomorrow."

"What papers?" Hassan questioned.

Mike told him about the restaurant he was planning to open and asked him to be his partner.

"Anything you want, big Mike," said Hassan.

"You know, no one is allowed to call me that except you," Mike shook his head smiling.

"Can I call you, Mike?" asked Andrew.

Mike nodded. Andrew continued, "I have a question. How do you know Hassan?"

"Not the time for this question," Hassan interrupted.

Mike turned to Andrew and said, "Why not let the kid ask. I am not embarrassed. Look, I forgot your name!"

"Andrew."

"Look, Andrew. The short version is; I was a very successful man until my wife died, after which I started drinking and gambling. I lost all I had earned and became homeless. My kids abandoned me. I used to come to this restaurant every day to get food. Hassan noticed me and asked me if I could work for him. He also offered me a room next to the restaurant if I accept the job," A tear fell on his cheek, and he continued, "I took the job, and we became close. One day Hassan asked me if I am ready to share my story, and I am glad I did. At that time, he was struggling to manage the restaurant. I helped him and taught him to manage everything. After a while, he came to me with a big cheque. He said this was my commission to teach him. Then he offered me a loan to start my own business. He talked me into it, and I took the loan, and here I am again. I lost a family, but I got him. Now, kid, you can figure out the rest," Mike pat Andrew on his back.

Hassan put his head on Mike's shoulder. Andrew looked at Hassan suspiciously, asking, who are you? What are you? Andrew couldn't figure out Hassan's personality, not yet, but that's not important. The thing that mattered was that Hassan gave him a family, a place which he could call home. Everyone has a complicated life, they have their baggage and fears, and despite having them all, Hassan never failed to smile and protect his family. He was wondering what Hassan's real family looked like and many other questions came to his mind.

Everyone was enjoying the dinner when Andrew saw Patrick at a table with someone, "Excuse me. I will be back in a minute." Hassan saw him heading to Patrick. He watched him closely.

Andrew greeted Patrick, who told him that the man sitting next to him is his boyfriend, Dave. Andrew greeted him as well and took a leave after a while.

Andrew and Hassan shared a look as he sat back down on his chair. He looked at his plate and said, "Thank you, Hassan." Hassan looked at Andrew and smiled. Patrick noticed a different Andrew that day. He followed him with his eyes to his table. He kept looking at Andrew from time to time. After dinner, they were all talking except Andrew. Hassan asked him, "Is everything okay?" He looked at Hassan and smiled, then said, "Everything is great. I just want to dance."

Andrew went inside, asked the DJ to put a particular song, and to turn the volume up. He returned and asked everyone to join him. Everyone stood up and started dancing to the song Andrew had chosen. Patrick was watching Andrew dancing with his friends. They were dancing, and many others customers joined them. Everyone had fun. It was a fantastic evening where they all laughed, enjoyed, and got even more closer. After an hour of dancing, they all decided to head home.

While everyone was talking over each other, Andrew looked at Hassan and thanked him. Hassan told him he didn't do anything, but Andrew said it was Hassan who made him strong enough to face Patrick without breaking. They both hugged and appreciated each other.

Marguerite and Mike had to leave, so John dropped Marguerite and Andrew took Mike.

On their way home, Mike and Andrew got to talk more and Mike confessed a secret, "I knew you are a doctor."

Andrew nodded.

"Can I ask you something?" Mike questioned.

Andrew nodded again.

"I want you to take care of my Hassan after I am gone," said Mike.

"What do you mean?" Andrew looked at him suspiciously.

"I am dying," Andrew stopped the car and listened to Mike. He was holding back his tears. Andrew looked at Mike with tears in his eyes, "I promise you I will take care of him. And your secret is safe with me," said Andrew assuringly. Andrew dropped Mike off and came back. John's car wasn't in the driveway, so he waited for him outside. John reached home and saw Andrew waiting for him. which made him realize it's now time to talk.

As soon as John sat beside Andrew, he asked, "Can you tell me why you are cheating on Hassan?"

"I am not cheating on him, " John said.

"You are telling me, you never cheat on him," Andrew questioned.

"I didn't say this," John looked at Andrew who was staring at him. Andrew arched his brows and let John continues, "I did cheat on him in the past but I have my reasons."

"What exactly these reasons were?" Andrew angrily stared at John.

"I received pictures of Hassan naked with someone," John responded.

"If he was cheating on you, why would he take pictures?" Andrew questioned.

John hesitated for a while then showed him the picture. Andrew looked closely at the pictures and said, "I was right. Hassan has a nice furry ass." John hit Andrew and took the phone, "I told you he cheated on me."

"Sorry to say this, but I'm not quite sure if you are dumber than you look or your jealousy has blinded you. Either way, you're mistaken," said Andrew.

"What do you mean?" John looked confused.

"Give me your phone," Andrew took the phone and zoom in on Hassan's hand. "Can you tell me what does that mean?" Andrew said.

"This is a hospital bracelet," answered John.

"Do you remember when Hassan had this?" Andrew asked.

"When he had the clots," said John.

"Now, can you tell me where you were?" Andrew further asked.

"I left for Montreal for work on that day," said John

"What time did you leave?" Andrew asked.

"Early morning," John answered.

"Look at the clock next to the bed," Andrew said while pointing at the phone.

John zoomed in at the clock and realized it's only 17 minutes after he had left.

"Do you remember if Hassan was under any medication?" Andrew asked.

"Oh, my God," John said, realizing how dumb he had been all this time. He understood the point Andrew was making that someone took advantage of Hassan. He started crying.

"You need to tell him," said Andrew.

"I can't, and if you tell him, he will leave me. Please don't break us. I can't live without him. This will destroy us," John pleaded.

Andrew didn't want to hide something this big, but he gave in and said, "Remember, if it comes down to picking aside, I will be on his."

Andrew continued, "Wipe your tears and go to him. You don't know how lucky you are, and don't beat yourself up. You were a victim of a mastermind plan. But remember, if you ever hurt him again, I will hunt you down."

"Believe me, I won't," John assured.

John and Andrew went into their rooms. John was upset, so Hassan consoled him and asked if everything was alright. John apologized to him and said he would never hurt him again. Hassan held John between his arms and kissed him on the forehead, and said, "You are here, and that's all that matters."

Andrew couldn't sleep that night. He was flipping in his bed. He got up and walked out of his room. He looked at the atelier door. He walked toward it and, as he entered the room, he could see the damage that John had done. He found the burned notebook and started reading it from the last page.

"You are here, but you are not with me. It feels like a dark cloud has been passed on in our relationship. Your hands dead in mine. I am with you, but only your shadow is here. I am watching you disappear, leaving me alone. Loneliness is nothing I want. Loneliness is shut-off walls, locked windows, and doors. Without you, I am fading away. Without you, I cannot stay. The black cloud took everything from me, took everything I had. Now the clouds crying for both of us."

Andrew gasped and looked at the paintings on the wall, "Your life is so complicated. I cannot imagine what you have been through." He found a painting of Mike as he looked through the stack and said to it as if Mike was in front of him. "I will keep my promise." He left the room along with the painting. He went back to his room. When he checked his phone, there was a message from Patrick. Patrick had confessed he couldn't stop thinking about him and wanted to meet Andrew again, to which he replied that they could meet tomorrow.

He finally fell asleep, but even then, he couldn't keep Hassan off his mind as he had a dream about him, and upon waking up, he realized he has started to understand Hassan a little well.

In the morning, everyone was having breakfast that Pete had prepared for them. They were all enjoying and talking except Hassan. Everyone looked at Hassan, who was singing alone. Andrew asked, "Is there anything you want to share with us?"

"Sorry, its' a song stuck in my mind," Hassan smiled.

They all asked him to sing, but he agreed to play it for them. It was an Arabic-French song about happiness and a ray of hope within.

Kyle asked, "Can I ask you why is this song stuck in your mind?"

"I don't know," Hassan answered.

"Leave him be," John said, walking towards them.

"Hey, boyfriend, no one asked you to defend him," Andrew teased.

Andrew continued, "He made us eat those salty eggs for breakfast that John prepared. It seemed like he cooked it with tears."

John was drinking water, and the water came out from his nose. Everyone was laughing, and Hassan gave him a tissue.

"It seems you had fun at breakfast," Kyle asked.

"You have no idea," Pete replied.

"You should move here so you will not miss the fun," Cara winked at him.

Kyle gave Cara a look, and they all continued to talk and laugh for a bit. Andrew then said, "Hassan, today, I will take you to Mike's."

John offered to take Hassan, but then Andrew looked at John, and he understood that something was up. "Okay then. Andrew will take you," John said.

Hassan nodded and put his hands on his head. He looked pale.

"Are you alright?" asked John.

"It just a headache," Hassan assured.

Andrew put his hand on Hassan's forehead, "You don't have a fever. That's good."

"He drank a lot of wine last night," said John.

"Maybe it's the eggs you ate," Andrew teased and continued, "anyway, drink lots of water, and you will be better."

Hassan nodded.

"I know the right treatment for your headache," John smirked.

"Get a room," they all said.

"After breakfast, you can have some dessert," Pete winked.

"Can I get some dessert too?" Ethan looked at Pete.

"Hey, we are eating," said Kyle and then whispered to Cara, "Can we get some?"

Andrew smiled, looking at them all happily in love, and asked them to stay the same. He knew he would find love someday too.

Kyle then turned to Cara and whispered, "Did you mean it when you said I should move here?"

Cara looked at Kyle and bit her lower lip, then she said, "I meant it."

"Guys, do you mind if I stay here for a while?" asked Kyle.

"Don't ask a question. You know the answer already," Hassan sighed.

Andrew interrupted, "Let me translate it for you. You are a part of the family, and your girlfriend lives here. So, you don't need to ask. This is the polite version."

"What is the other version?" Kyle asked.

"Shut up and do it already," Andrew responded, and they all laughed.

After breakfast, Hassan and Andrew went to get dressed, leaving Ethan and Kyle on dishes duty. Pete was talking to Cara about the work. John came from behind, kissed Hassan and told him about the painter. Hassan asked him to get everything before the painter arrived and said he wants the atelier to be painted light blue.

John asked, "Are you okay after what I did?"

"I have moved on, and you are still here," Hassan replied.

"I don't know what to say," John was ashamed.

"I love you," Hassan smiled.

"I love you too, shorty," John sighed.

"See, things are back to normal. You said shorty," Hassan smiled and kissed John. He then changed and left the room with John.

Andrew asked Hassan to wait for him in the car. As Hassan left, John thanked Andrew, who had asked him to be good to Hassan.

Andrew and Hassan left the house. While driving, Hassan saw a baby with his father crossing the street. Andrew noticed Hassan was looking at the baby and asked, "Are you alright?"

"John asked me to adopt a kid after getting married," Hassan replied.

Hassan looked tense, and Andrew comforted him, saying, "You are a family person with a very task-oriented personality. You would be a terrific father and parent."

"You think so?" Hassan questioned.

"I wish I had a father like you; I'd be a very different person right now. I don't know what kind of family raised you, but you take care of everyone around you; you would be a wonderful father, any kid would be lucky to have you." Andrew responded.

Hassan thanked him, and they entered the new restaurant to meet Mike.

They greeted Mike, and Hassan said, "This place is impressive."

"I am glad you like it. Now, I need you to sign the contract," Mike handed him the papers.

Hassan started reading, but Mike and Andrew stopped him and asked him to trust Mike and sign. Hassan signed the papers. Andrew didn't want Hassan to find out what the papers were.

After Hassan finished signing the papers, Andrew took them, put them in an envelope, and returned them to Mike. Mike gave the envelope to the lawyer.

"Did you think of a name for it?" Hassan asked.

"Nie?" Mike replied.

"No, it's your restaurant. What do you think of Big Mike?" Hassan suggested.

Mike rejected it instantly, then Andrew said, "I have a suggestion. Why don't you name it Mike N' Nie?"

They both agreed. Andrew gave Hassan the keys and asked him to wait in the car. Hassan left, but he couldn't help but think, "Why was Andrew acting weird?" Andrew looked at Mike and smiled, "Take care of yourself." Mike shook hands with Andrew and said, "Thank you, kid. Take care of him." Andrew smiled and walked away. A warm tear dropped from his eye, knowing it could be the last time to see Mike. He was thinking about Hassan, how it would affect him. He got in the car and drove home. Hassan was a little angry. Andrew looked at him and said, "It's fine; just say it."

"What happened there?" Hassan asked.

"To be honest, if you read it or not, will that affect your decision?" Andrew explained. Hassan knew he was right. Andrew then confessed, "I feel I am mature now. Before, I wasn't like this. I have always been weak and I evaded every crisis that came my way. But now, I see things clearly and do things I feel right doing without feeling the need to prove anybody. I am no longer blinded by 'love.'"

"You are not weak, never have, never will be. You just had a hard time because you thought he was the one. Wait, where are you going?" Hassan screamed.

"You will know, Hassony," Andrew replied. Hassan's eyes filled with tears and he asked, "What did you just call me?" Andrew sat on the park bench and answered, "Hassony, that what I called you."

"I am not crazy," Hassan cried.

Andrew hugged him and told him he is not. Andrew cried after he hugged Hassan. Then Andrew looked at Hassan's face, wiped his tears. "You are the strongest, tiniest person I ever met." Hassan laughed with tears in his eyes. Andrew looked at Hassan's face and said, "Remember, I am always here for you." Hassan threw himself on Andrew's chest and cried. Andrew wiped his tears and hugged Hassan for a while. Andrew asked Hassan if he would come with him to meet Patrick and Hassan agreed. Andrew asked Patrick to come to the park. They waited in the car till he arrived. Andrew asked Hassan to wait in the car. He then got out of the car and walked towards Patrick.

Patrick greeted Andrew and asked, "Is he your boyfriend?" he pointed towards the car.

"No, he isn't; he is family. He already has a boyfriend," Andrew answered.

"Good. I was thinking of getting back together if that possible," Patrick suggested.

"Look, Patrick, If I want to be with someone, I'd prefer someone like him," Andrew pointed towards the car and continued, "not with someone like you. He is very supportive, sensible, and kind. You ruined my life, destroyed me. It took me two years to realize that you were the wrong guy for me. When I think of it, I never asked you for much. I just wanted your love, respect, and appreciation, and you couldn't even give me that. In comparison, I did a lot for you. I always had your back; I supported you and pushed you forward. But now I am done! Find someone else to play with his feelings, and remember, one day, you will reap what you sow. Don't ever contact me again."

Andrew walked to his car, started the engine, and drove home. Hassan was looking at him but didn't say anything. After a while, Andrew took a deep breath and looked at Hassan, and smiled. Hassan asked what happened there, and he told him the complete story. He said, "I lectured him and said a big fat no. I want to dance, yell and do many crazy things. I feel a giant rock was on my heart, and I just moved it. I told him he was the wrong guy for me.". Andrew thanked Hassan for all the things he had done for him.

After 30 minutes, they arrived home. Andrew parked the car, and they walked together to the door. John opened the door and said, "You are back." Andrew opened his hand and took a deep breath, "Yeah, it was an amazing ride." John looked at him, then asked Hassan, "Is he high?"

Hassan told John about how Andrew boom slapped his ex in the face. John laughed and said, "I want to do boom, but here," he touched Hassan's ass.

Hassan put his finger on John's nose and said, "You dirty man!"

"I can be dirtier," John winked, and they went inside holding hands.

Andrew saw Ethan was sitting in front of the TV. He went to him and apologized if he had ever hurt him. Ethan told him he never did.

Thirty-five days of love and happiness passed. Everyone at that house was waiting for this day, the reason they are together. It was Hassan's birthday. They all were busy preparing for the birthday party. It was the end of the workday. Hassan received a call and left everything behind. The call disturbed him; he called an Uber. Uber driver saw him worried and asked if he was alright. Hassan broke into tears but didn't say a word and got off the Uber as soon as he reached the destination.

Everyone at home was waiting and worried as Hassan was late and not answering his phone. As soon as Hassan arrived, John informed everybody, and they waited for him to open the door to surprise him. He opened the door, and everyone yelled, "Surprise." Everyone was quiet and shocked when they saw him crying. Andrew moved close to him and asked him, "What happened?" Hassan looked at Andrew was crying and said, "Mike." Andrew understood that Mike had passed away. He hugged and walked him to John. John tried to calm down Hassan, but he couldn't stop crying. Everyone was heartbroken over the sudden reveal. It was a sorrowful moment.

Andrew asked John to take everyone to another room as he wanted to talk to Hassan privately. He looked at Hassan, wiped his tears with his hand, "You are torturing him. He wanted you to be happy. He was sick and suffering but couldn't tell you. At least he is in a better place, now." Hassan stopped crying and put his head on Andrew's chest and closed his eyes, and said, "It hurts because it mattered, and it will always matter."

Andrew consoled, "Love will always leave a window open for fresh air in the middle of the storm. There's an invisible thread connecting your heart. I believe Mike will live in your heart forever. Now go get some rest."

Andrew called John to take him up to his room, "Don't leave him until he feels better." John took Hassan upstairs.

"You knew Mike was sick!" Cara asked. Andrew nodded and said, "Mike was very sick. He was afraid he would die before paying Hassan back as he built everything he had from Hassan's money. Also, his son was waiting for the moment he dies to take everything. So, he gave everything to Hassan. Hassan doesn't know what he signed. That will crack him again."

Hassan washed his face, and John took him to the bed. Hassan tried to close his eyes for a while. After a few minutes, Hassan opened his eyes, stood up. John asked him, "Are you alright?" Hassan moved a few steps and stopped. He couldn't hear John or see anything. All he could see was his old home. John understood that Hassan must be reminded of his past. He tried to talk to him, but Hassan didn't respond to him. Hassan started panting, John tried to calm him down and help him breathe. John started crying and screamed very loud, "Somebody help me." Everyone heard John screaming. They came running to John and Hassan's bedroom. They saw John holding Hassan and crying, "Please help me."

Andrew moved fast and held Hassan and tried to get him up. Hassan raised his hand like he saw something. Andrew moved him to the bed, held Hassan's head between his hands, and talked to him, "Hassan, listen to me," but nothing happened. He closed his eyes and calmed himself and said, "Hassony, look at me." Hassan moved his eyes and looked at Andrew and said with a weak voice, "It hurts. I cannot do this anymore." Andrew looked at him and said, "I know it hurts, and it causes unbearable pain. We are all here and we need you."

Hassan put his head on Andrew's chest and said, "I am sorry."

Andrew said, "Don't be sorry. We need you to fight for us." Andrew held Hassan between his arms and said, "Take everyone out of this room. I will stay here with him."

John insisted on staying, but Andrew convinced him that it would only make the situation worse. Andrew then looked at Hassan and said, "It's time to talk about all the sad things that happened to you in the past."

Hassan then said, "I come from a wealthy family. I had the best father in the whole world. He was a Muslim, but he never abandoned me because I was gay. He was very proud of me. I had a younger sister and brother and an amazing mother. We were a very happy family until one day, my grandfather showed. He did many horrible things to my family. My father was worried about dying before my grandfather because a legitimate blood relationship to the deceased is entitled to inherit according to Islamic law. To avoid this problem, my father transferred everything under my name. This enraged my grandfather, who took my father, and the next day my father returned in a box. I couldn't have the chance to say goodbye to him. They took him away from us." He cried and continued,

"At that time, I had a boyfriend. My family loved him. A year passed, but nothing changed. They kidnapped, tortured, and raped me with my boyfriend. I watched him slaughtered in front of my eyes. In one year, I've lost the two most important men in my life. I managed to escape, and when I returned, I found my family under the wreckage nothing left to me. Everyone leaves me eventually," he broke into tears.

Andrew embraced him and said, "No one wants to leave you. We won't give up on you, so please don't give up on yourself." Hassan then took out a small box from his closet and showed Andrew a picture of his sister, his dad's watch, his mom's journal, and his brother's bracelet he made for his birthday. The only memories he had of his family.

Andrew then said, "I know you miss them, but living in the past won't be right for you. We are your family. Share your moments with us." He then went downstairs and explained that Hassan needs time to process what happened today. Then he asked John to go to his room and act naturally, even make love if he could. John ran to his room. He opened the door and walked to the bed. He laid on the bed next to Hassan, hugged him, and kissed him, "I love you." Hassan smiled then said, "I love you too.". John kissed Hassan, and they both slept peacefully.

Chapter 10: María

Love is trust

"Have enough courage to trust love one more time and always one more time."

— Maya Angelou

A month later, Andrew and Hassan went together to complete some papers. After they finished, they stopped in a shopping center to grab a drink. They lined up after giving their orders. Hassan was talking to Andrew when a woman accidentally hit him and spilled a hot coffee on his shirt.

"Are you alright?" Hassan asked.

"I am sorry, I ruined your shirt," the woman replied, her hands were shaking.

"Forget about the shirt. Let us have a seat. Andrew, can you please bring a coffee with you. What is your coffee?" Hassan turned to the woman and asked. "Double-double large, please," she requested.

Andrew nodded. Hassan sat with the woman at a table. He held her hands and asked her, "Are you alright?"

"I am fine," she started crying.

"May I have your name?" Hassan asked.

"María," she replied.

"I am Hassan. María, you don't look fine," Hassan showed concern.

"I came here to see my son, but he has no time for me now. He won't talk to me," María broke down.

Hassan wiped her tears and kissed her on the forehead, "I know it's hard for you and painful, but at this point, you need to give him space. I am sure he will come back to his mother."

"Here is your coffee, and Hassan, this is your smoothie," Andrew handed them their drinks.

"Are you both a couple?" she asked.

"He wishes!" Andrew teased.

Hassan laughed and said, "No, we are close friends, more like a family. I have a boyfriend," he then whispered in María's ear, "and he is single. Maybe you can find someone for him. He is a little coco."

"I can hear you. I am not deaf," Andrew said annoyingly.

María smiled, watching both of them talking. Andrew looked at María and smiled, "And here is the big wonderful smile, show us more teeth."

"You both are a real family. Don't ever give up on each other," she said.

They nodded.

"I wish my son could learn from you," María sighed.

"He will come back to you. I am sure," Hassan assured her.

Andrew turned to María and asked, "I didn't get your name."

"María," she replied

"Wow, what a lovely name. Nice to meet you, María. I am Andrew," Andrew shook her hand.

"Nice to meet you too, Andrew," she shook back.

"We call him Dewey," Hassan mocked.

"I hate that," Andrew groaned.

"Why do you call him that?" she asked.

"For an Andrew, that's gentle as the morning dew," Hassan explained.

"That is nice," María smiled.

"See...," Hassan turned to Andrew.

"Do you know what his boyfriend calls him?" Andrew asked María.

"What?" she replied

"Shorty, what kind of nickname is that?" Andrew laughed.

Hassan stuck his tongue out and said, "He can call me whatever he wants."

Andrew started mocking Hassan. María was smiling, then laughed, "I can't believe I am laughing right now. You both made my day."

"Laughter is the best medicine. It draws people together in ways that trigger healthy physical and emotional changes in the body," Andrew explained.

"Not now talking encyclopedia," Hassan teased.

María looked at both of them and said, "Andrew, I hope you will find someone who will love you the way you deserve. Hassan, I wish you happiness, and I hope Mr. Nicholaus will find someone like you."

"Who is Nicholaus?" Andrew asked.

I work as a housekeeper and a nanny for Mr. Nicholaus. He is a wonderful guy," she answered.

"Why did you wish him to find someone like Hassan and not me?" Andrew questioned.

"Because he would be a perfect fit for him. He is a family guy already, and you are halfway there," she patted Andrew's head and continued, "It was wonderful to meet you both, but now I have to go."

María stood up and said, "Good luck for you both."

Hassan and Andrew said in sync, "Take care of yourself. Goodbye."

María left the table and walked away. All she was thinking if ever she felt down, she will think about Andrew and Hassan. She turned to look at them from far and wished that Mr. Nicholaus will meet someone like Hassan, who has the same family values. She smiled and walked away.

"She was right. You already there, and I wish to find someone like you too—someone who will support me and push me to be a better person. I am jealous of you and John. I never had such a connection," Andrew sighed.

"You will find someone. You will have it all," Hassan assured.

"That's what Nie said to me," Andrew added.

"If he said that means you will," Hassan stated.

"I still don't understand why did Nie choose me?" Andrew wondered

"It is simple. You are me," Hassan shrugged his shoulders.

"But taller and more handsome," Andrew chuckled.

Hassan laughed and said, "Time to go home."

"I won't walk with you with this hideous coffee stain on your shirt," Andrew stepped back.

"I will buy you ice cream," Hassan smiled.

"Let's go," Andrew and Hassan bought ice cream and ate it before driving home.

Seven months passed on, and the house was full of joy and happiness, warmth with love. Kyle moved to live with Cara. Andrew was doing good with his career and started dating an IT tech, Ang. Pete was sleeping with Ethan every night after they had a big fight. John and Hassan were very happy together. John was planning a trip for both of them. Hassan had a big opening for the new restaurant. Brian found a job at the same company Ethan and Hassan work with, but in a different department.

While Cara and Kyle were sleeping in their room, Cara had a nightmare about her deceased husband, and she woke up and screamed, "Dave." Kyle woke up with her scream, hugged her, and tried to comfort her. Hassan Knocked on the door, and Kyle said, "Come in." Kyle asked Hassan to talk to Cara, and he left the room. He couldn't hold his tears any longer, so he left the room and went downstairs.

"Did you have the same dream?" Hassan asked.

"Yes, I don't know what to do," Cara cried.

"It's your mind unconsciously sending you a continuous message. You are finally happy, but you are afraid to lose Kyle," Hassan wiped her tears. "Maybe you are right," she said.

"Don't do that to yourself. Kyle won't leave you, but if you continue like this, you might lose him," Hassan warned her.

"Please go and talk to him," she requested.

Hassan got up and said, "On my way to talk to him. Try to sleep."

Hassan went downstairs and found Kyle crying. Hassan put his hands on Kyle's shoulder and said, "You don't need to beat yourself."

"Is it the same dream?" Kyle sobbed.

Hassan nodded.

"She is still in love with him," Kyle cried a little louder.

Hasan hugged him and said, "She is in love with you. I will tell you what I have told her already. She is happy now, and that makes her terrified to lose you. Her mind in unconscious levels sends her a continuous message to be careful not to lose what she has."

"What does she have?" Kyle looked him in his eyes.

"You," Hassan replied without blinking.

"Do you think so?" Kyle was looking for answers.

Hassan kept one hand on his shoulders and said, "I am sure. Let me tell you something about Cara. After days, she promised herself she won't be with another man, but she is now with you. She has plans for the future with you. She wants to have three kids, one boy and two girls."

"She wants kids with me!" Kyle smiled.

Hassan nodded, "Wipe your tears and go to her and tell her that you are in love with her, and you won't leave her no matter what. Be there for her."

Kyle got up and thanked Hassan. Hassan told him he just wants both of them to be happy together.

Kyle went upstairs while Hassan stayed. Kyle entered the room, laid on the bed next to Cara, and told her, "I love you, and I am here. I will always be here with you." Cara threw herself on Kyle's chest and cried, "I love you too, I do." Kyle wiped Cara's tears and kissed her passionately. Kyle wrapped his arms around Cara, stayed like this until they fell asleep.

John came down and asked, "Are you alright? Why didn't you come to bed after?"

"I am fine. I am just worried about them," Hassan replied.

"You should be worried about me. Tomorrow I have a big meeting, and I didn't even have time to pick up a suit," John teased.

Hassan got up, "Alright, let's find you a dress."

John went with Hassan to their room. Hassan opened the closet, took a blue suit, and said, "I picked this for you and steamed it for you.".

"That's why I love you." John said.

John tried the suit on after Hassan picked a light blue shirt with it. Hassan then handed him a tie.

"Woah, from where you got this tie," John asked.

"I searched a lot until I found a red wine color tie, floral on one side," Hassan responded.

John took a deep breath, "I love you."

"I know," Hassan smiled.

John changed and said, "I will change, and I will need some stress releaser."

"You are a bad boy," Hassan chuckled.

John undressed himself. And came to the bed naked and pulled Hassan from his legs. He looked at Hassan and asked him, "Why do you still wearing pajamas?". Hassan looked at John and bit his lip, and whispered in John's ear, "Waiting for you." John undressed him, and then they both made love till they fell asleep.

While sleeping, Hassan had a nightmare. He couldn't breathe, then the clock alarm woke him.

"Morning, shorty," John kissed his forehead.

"Morning," said Hassan.

"Do you want to join me in the shower," John asked, but Hassan told him he needs to sleep more.

John kissed Hassan and went to the shower. After a few minutes, Hassan got up from the bed and went to the shower to join John. Hassan entered the shower and grabbed John's ass. John bent forwards and said, "You know what I like." Hassan spanked John's ass until it turned red. Then he buried his face between John's butt cheeks. John stood up, took Hassan out of the shower, dried Hassan, and took Hassan to the bed. He was on top of Hassan. He felt Hassan inside him, and then he moved up and down. Hassan moved fingers on John's legs, and that drove John crazy. He kept moving even after Hassan finished. John stood up on the bed and sat on his knees. He placed Hassan's legs on Hassan's chest pushed himself inside Hassan like a wild tiger. He wrapped his hand around Hassan and started moving with long deep passionate kisses. He moved and put a pillow under Hassan's back and kept moving. Hassan grabbed the sheets with his hands, and his moan got a little louder. John lost control and went a little rough. John and Hassan were sweating from every pore of their skin. John leaned forward, held Hassan between his arms, and kept moving.

John was growling aloud. After he finished, he began to kiss Hassan, kiss after kiss like he never kissed him before. It is like an endless love drove them to melt in each other, took them to heights, and made them running in an infinite loop to each other. Kiss after kiss to tell the whole world they meant to each other. The touching drew a storyline of memorable moments and to connect emotionally to the highest level. All of them together means love because love is not words to say. It is life, share, style, and giving. You can learn thousands of words about love, but all words in the world cannot express your love for someone without creating your own language to express that rush in your vein when you are with the one you want to be with.

Later, John and Hassan joined the rest of the family for breakfast. Everyone was looking at John and Hassan. There was something different about them, but no one could figure it out.

"You both look different," Andrew asked.

"I am wearing a new suit," John responded.

"No, not that. You are both have a kind of glowing," Pete added.

Hassan and John smiled and shrugged them off.

"Woah, this is a lovely tie from where you got it?" Kyle pointed at John's tie.

Andrew jumped in and said, "No way this tie and suit is your taste," Andrew looked at Hassan, "You did this, am I right?"

"I picked a suit for my boyfriend. It's something a boyfriend does," Hassan smiled.

"Damn you. Next time you will pick one for me," Andrew insisted.

"Ask your boyfriend," Hassan teased.

Andrew rolled his eyes and said, "Technically, we are not there yet, and honestly, I am getting bored."

Andrew pointed his fingers to Hassan, "I will curse you if you didn't come with me shopping."

"What a drama. Fine," Hassan smiled.

Everyone was laughing. They enjoyed the breakfast. After breakfast, everyone went to work except Pete and Hassan. Hassan asked Pete to bring something from the grocery shop. Pete left his phone at home. Pete's phone was ringing. Hassan took a look at it and answered. The call was from a debt collection agency. Hassan talked to the agent, collected all the information he needed, and recorded it on paper. After the call, he kept looking at the paper. Pete returned with what Hassan asked him to buy. When Pete entered the kitchen, Hassan kept looking at him with a weird look. "Is something wrong?" Pete arched his brows.

"The debt collection agency called," Hassan handed him the paper.

"You don't know the story behind it," said Pete.

"I figure it out, but how can I have missed it. And why you didn't tell me?" Hassan asked.

"It is my problem, and I have to deal with it," Pete said firmly.

"With all respect, how do you want to deal with it?" Hassan questioned.

"I will figure out something," Pete started crying.

Pete moved closer to Hassan. Hassan hugged him, "You don't need to stress yourself. I can help you."

"It's not your fault. You don't have to," Pete said.

"Not your fault either, and you were too young to handle all of that. I will help you. I will pay it on your behalf, and you will pay me back. From today, you won't pay me the rent. Is that okay?" Hassan proposed.

Pete looked into his eyes and thought, how can someone be this selfless, "This is too much. I don't know how to thank you."

"By taking care of yourself," Hassan hugged Pete.

John arrived at work, and he was very nervous. He was prepared but still anxious. He met his boss at the elevator.

"Good morning," John greeted.

John's boss, Allan, greeted him back. "Oh my! This is a lovely tie," Allan added.

"Thank you, sir. My boyfriend picked it for me," John admitted.

"It seems your boyfriend has good taste," Allan said.

"Yes, he has. He picked me," John smirked.

"See you at the meeting. Have a good one," Allan said and leave.

John walked out of the elevator. He took his phone from his pocket and called Hassan. Hassan answered right away, "I knew you would call. What's up?"

"I think I will puke," John said anxiously.

"Sweetie, you are prepared for this moment. I know you will be a Rock star, and everything will be great. Then you will come home to talk about it, and we will do some of your sexual fantasies. We will have a fantastic night together. " Hassan responded.

"How dirty are we talking?" John laughed.

"Use your imagination," Hassan smiled back.

"I love you. See you at home," John ended the call and said to himself, "You can do this." and walked to his desk.

Pete was next to Hassan and heard everything, "What was that?"

"John called from his work," Hassan answered.

"I know that, but what is the part of sexual fantasies. Do you guys have sexual fantasies?" Pete inquired.

"Everyone does. Don't you?" Hassan shrugged his shoulders.

"I didn't try mine yet," Pete sighed.

"I am sure Ethan will be open to trying some of them," Hassan responded.

Pete remained silent and then blurted, "I want to see other people."

"What? Ethan isn't enough for you, or are you looking to try new things?" Hassan stared at him.

"Since I arrived in the city, I've been with just Ethan. Is it wrong to think like that?" Pete asked.

"It's not wrong, but are you willing to risk the things you have just to try something?" Hassan questioned.

Pete looked confused, "I just want to try."

"Talk to Ethan about it. I am sure he will understand. But be careful," Hassan suggested.

After hours, John finished the meeting. The meeting was successful. All were impressed by John's presentation and ideas. The other party suggested having a business dinner at a restaurant.

Another company's employee, Christopher, suggested a new restaurant that his daughter told him about. On hearing the name, John smiled because it was Mike N Nie. John assured them he would make the reservations for tonight's dinner as it is his boyfriend's restaurant.

John walked out of the meeting room after everyone left the room. John called Hassan.

"How was the meeting?" Hassan asked.

"The meeting was great. There is something I need to tell you. They suggested having a business dinner at a restaurant, at your restaurant," John specified.

"That not even a problem," Hassan responded.

"Oh, thank you—one more thing. My boss wants to meet you," John asked.

"I will be there. Text me the time and the number of guests. Also, don't let your boss pay, and you will tip the waiters after, Hassan asked.

"No, that's a lot," John hesitated.

"Do as I said. You won't regret it," Hassan smiled.

"I love you," John said

"I love you too," Hassan replied and dropped a line.

Hassan asked Pete to go with him to his room. Pete and Hassan went to Hassan's room. Hassan opened his closet and picked up a new shirt for Pete to wear.

"I think this will fit you."

Pete stared at him and said, "They want to meet you, not me."

"Meeting me includes meeting my family. You all need to prepare. We are going to set everything up," Hasan handed him the shirt.

Pete left the room to prepare. Hassan took his phone and texted everyone to come to the restaurant. Pete and Hassan went to the restaurant together and met everyone there. John, his co-workers, and the others arrived, and they found everything was set up for them. They sat, and Hassan came with Pete to their table. Hassan introduced himself and Pete. The rest came after him, and he introduced every one of them. Everyone ordered what they liked to try. They talked during the time until dinner has been served. Hassan didn't join them but observed everything closely. Everyone was enjoying the food, the service, and the restaurant's atmosphere. John saw Hassan walking, wearing a suit, and he sailed away with his thought. He imagined Hassan wearing a wedding tuxedo, holding blue flowers in his hand, looking at him. The moment that John waited so long. He looked around, and everything was moving in slow motion. John smiled and looked at Hassan. Everyone looked at him then looked at Hassan. Suddenly he heard, "Someone is deeply in love."

"Sorry," John blushed.

"You can't stop looking at your boyfriend. For how long have you been together?" Christopher asked.

"Almost six years," John replied.

"And you yet can't stop looking at him! You must deeply in love with him," Christopher smiled.

"I am," John smiled.

Christopher raised his glass, "To John and his boyfriend."

Everyone raised his glass and wished a happy life for the happy couple. John left the table for a few moments. He held Hassan from behind. Hassan turned and looked at John, and then John kissed him, "Thank you." John returned to the table. Hassan kept looking at John with a smile on his face. "You both couldn't wait until everyone has gone to kiss," Andrew said while standing behind Hassan. Hassan turned around and said, "Stop showing up like this. You will kill me one day with a heart attack."

Cara asked Andrew about his boyfriend. On which Andrew told them that they broke up.

"Shocker!" Ethan mocked.

"Have you considered to stop mocking him?" Kyle suggested.

"If he cannot handle that, why are we together? I am a catch," Andrew said.

Everyone was laughing at Andrew's over-dramatic performance.

After diner, Allan came to Hassan and asked him if he could do something special for his anniversary next week. Hassan told him that he would do anything he can to make it a memorable anniversary. Allan thanked Hassan for the lovely evening then left as everyone left.

They all went home. John and Hassan went to their room. Cara with Kyle, Pete with Ethan, and Andrew alone. John changed and laid on the bed next to Hassan. He held Hassan between his arms and kissed him. Hassan looked at him and smiled.

"What?" John asked.

"Just want to figure out what you about to do?" Hasan replied

"I don't need any fantasies anymore. You are my fantasy. You are all I want," John kissed his forehead.

Hassan put his head on John's chest and listened to his heart beating. Then they heard someone was knocking on the door. Hassan got up and opened the door. It was Ethan. He wanted to talk to Hassan in private. They went downstairs to talk. Ethan was upset, "How could you do this to me? What did you tell Pete today?"

"He told me what he was thinking, and I told him to talk to you. Am I missing something?" Hassan was worried.

A tear rolled from Ethan's eyes, "I loved you for years, and now I am happy, and you encouraged him to explore his options. I don't want to lose him."

"You are falling for him," Hassan hugged him and continued, "I asked him if he is willing to risk everything he has, and he wants to take the risk. He made up his mind."

You are telling me to give him up," Ethan stepped back.

"I didn't say that. Please listen carefully to me. Let him try, but you have to do something. Love is like a game, don't be in the game unless you are ready to catch him. If you really want him, you need to listen to me," Hassan explained everything to Ethan about what he has to do. Ethan agreed with the idea. Ethan went downstairs and told Pete that he agreed with his decision, and they can see other people, and if Pete wanted to be with him, he would be more than happy.

After Ethan talked to Pete, he left his room and closed his door. Andrew was about to enter his room and saw Ethan walking. He followed him and saw him crying alone. Andrew sat behind him and held him between his arms, "I don't want to lose him."

"You won't lose him, Andrew assured and continued, "I didn't realize you are in love with him. I thought you filled that space in your heart. I thought you weren't serious, but now I can see how much you love him."

"I loved Hassan for years, but this is different. I don't know what to do," Ethan looked helpless.

"You will be with him. I promise you. But now get some rest," Andrew asked.

Ethan went to his bed, tried to sleep. Andrew was thinking about how he could help Ethan. He went to his room, laid on his bed, and fell asleep. He saw a dream where all of them were together, hugging. It felt real, but as he opened his eyes, he found himself in his bed.

Chapter 11: Unforeseen Trade-Offs

An unforgivable mistake!

"There is a saying in Tibetan, Tragedy should be utilized as a source of strength.' No matter what sort of difficulties, how painful experience is, if we lose our hope, that's our real disaster." — Dalai Lama

In the morning, everyone was sitting at the table, having breakfast. Ethan didn't sit next to Pete as usual. Hassan looked at everyone and said, "Why a dreadful silence?".

Andrew answered, "It seems not everyone is in the mood to talk."

"I didn't get enough sleep," Cara replied. Kyle nodded in agreement.

"I thought we talked about it," Hassan looked confused.

"We are trying to figure things out," Kyle explained.

"Stop wasting time. You are deeply in love with her, and she loves you. Everything else doesn't matter. So, stop acting like babies and try to be happy together. You are building a life together," Hassan then turned to the other side, "Pete and Ethan, you both need to have an honest long grown-up conversation to figure out what you both want. And you, Andrew, I don't know what is wrong with you, but find a hobby or a boyfriend. The sourpuss doesn't suit you. And John, today we have the annual blood test." Hassan lectured, and nobody argued.

"I'm not too fond of needles," John hesitated.

"You mean you are afraid of them," Cara added.

John nodded.

"Why are you afraid of needles?" Kyle asked.

"He has an extreme fear of medical procedures involving injections or hypodermic needles. It is called Trypanophobia," Andrew explained.

"Thank you for giving us the medical term, Mr. talking encyclopedia," Ethan mocked.

"Don't mind him. He just a little moody this morning," Pete said.

"Are you both okay?" Andrew asked.

"I am okay," Ethan and Pete said in unison.

Kyle showed concerned, and they both said they are figuring out some issues just like Cara and Kyle. "I don't need to figure things out. She is the most amazing woman I ever met, and I love her deeply," Kyle responded. Everybody was in awe to hear such beautiful words.

John then reminded Hassan about the Europe trip and request him to do his packing too, and Hassan agreed.

"For how long will you be gone?" Cara asked.

"Two weeks," John replied

"Two weeks!" Pete repeated.

"After what they have been through, two weeks would be an excellent time to relax," Kyle added.

"You are both lucky," Andrew said.

"You both deserve to be happy," Kyle added.

"We will miss you both," Pete joined the wagon. The whole group gathered and did a group hug.

After breakfast, everyone went to work. John and Hassan went to the lab to do the routine blood work. After hours, Hassan received a call from his doctor to redo the blood work again because it was a mixed-up between lab results. Hassan informed him that they would do it tomorrow.

The next day, John and Hassan went to the lab. Hassan noticed that there were more tubes than before. Hassan asked the technician, and the technician said that the doctor ordered more tests and no need to be worried; it just a routine. Hassan contacted his doctor, who said it's just a new routine. Hassan agreed but wasn't convinced at all.

It's been a day since Hassan and John left, and everyone at home missed them already. Pete was leaving for his date, which made Ethan upset, and finally, Andrew found someone on a dating app. Cara and Kyle, on the other hand, were in total silence. Cara and Kyle were sitting on the couch. Kyle was switching from one channel to another. Kyle received a text message from Hassan saying, "Dance with her." Kyle smiled and switched to YouTube and picked a song. He took Cara's hand and danced with her. He wrapped his arm around her, and she put her head on his chest, "I need you, and I want you to need me back." Cara looked at Kyle and cried, "I am scared of losing you. I love you." Kyle smiled and said, "I am here. I won't leave you. I love you." Cara smiled and put her head on Kyle's chest, and continued dancing.

Pete waited outside his date's building until his date arrived to go somewhere. Steve was the name of his date. Steve asked Pete to come with him to change his clothes before going. Pete agreed and went with him. They entered the building and entered the elevator. At the elevator, Steve grabbed Pete hard from his private area and said, "I pet you like it rough with a roleplay. We will have so much fun." All the memory came back to Pete's mind like he was re-living the past but with a different person. The elevator's door opened; Steve stepped out of the elevator holding Pete's hand. Pete looked at Steve with a ferrous look and said, "I will never let anyone mistreat me again. I am a man, and if you can't handle that, I don't need you in my life."

"Now I am the bad guy!" Steve questioned.

Pete pulled his hand out and said, "That makes you the wrong guy."

Pete left the building, and he was thinking about how he got himself in this situation. He took the subway to go back home. He arrived home and went directly to Ethan's room and found Ethan with someone in the bed. Pete closed the door and stood behind the door, thinking, "What I have done."

He opened the door again and asked Ethan's date to leave.

"What are you doing?" Ethan asked.

"He needs to leave," Pete grabbed Ethan's date from hand.

"Who are you?" Ethan's date looked confused.

"I am his boyfriend, and now get out," Pete declared.

The date left, and Ethan asked, "Did I hear you said you are my boyfriend? How?"

"When I saw you with him, I felt I am about to lose you. It was an unbearable feeling. I am sorry I hurt you; I won't do this again, Pete grieved.

Seeing Pete upset, Ethan told him the truth about the fake date. It was Hassan's idea to make Pete realize what he is losing.

"Do you want to be with me?" Pete's eyes filled with tears.

Ethan sobbed, "I am falling for you. You said you are my boyfriend. Do you mean that?"

"More than anything," Pete smiled.

Ethan moved close to Pete and hugged him.

"I am tired. Can you cuddle me?" Pete said. Ethan put Pete on the bed and cuddled him. Pete closed his eyes and took a deep breath.

Andrew was enjoying his date, Shane. They spent the whole time talking. After the date, he returned home happy, singing. He went to Ethan's room to talk about it and found Ethan holding Pete between his arms, "This day is getting better."

"How was your day?" Pete asked.

"It was amazing. What about your date?" Andrew asked back.

"You are seeing me in Ethan's arms. What do you think?" Pete replied.

"I am sorry," Andrew said.

"Don't be. I am happy it didn't go well with my date. I already have a boyfriend," Pete smiled.

"You heard him. He is my boyfriend," Ethan repeated.

"You are both meant for each other. I am happy for you both. Can I lay on the bed and talk?" Andrew asked.

All three of them laid down. Andrew said, "I love this house, and I just realized something."

"Now, What?" Pete rolled his eyes.

"Ethan is the only one here who almost dated everyone in this house. He was in love with John and Hassan. He took me on a date, and now he is finally with a boyfriend," Andrew answered.

"Are you trying to sabotage my relationship?" Ethan slapped Andrew on his back.

"Ouch, you are totally in love with Pete. The point you were attracted to unique people," Andrew cleared.

"Are you saying you are unique?" Ethan asked.

"Yes, I am, Andrew chuckled and continued, "everyone in this house is unique."

"We are a family," Pete added.

"This is the most important part. Do you know John will propose to Hassan when they will come back from Europe?" Andrew broke the news.

"They deserve to be happy," Pete added.

"I hope everything will be as per their plan," Ethan wished.

"What do you mean?" Andrew questioned.

"He means every time John and Hassan were happy, something happens," Pete explained.

"Oh, God. I hope nothing will go wrong," Andrew prayed, and Pete and Ethan amen to that.

Fourteen days have passed, and everyone at home was waiting for them to arrive. Hassan texted Cara about an hour ago. Now they were about to arrive. Cara, Pete, Kyle, Ethan, Andrew, and Shane were waiting for them to arrive. Shane went to the washroom. After a minute, John entered the house with two suitcases. He dropped the suitcases and started dancing and shake his ass and was saying, "Oh yeah, I had sex on the beach."

"My ears," Cara covered her ears.

"We don't want to know," Ethan added.

"Leave him be. He is happy," Andrew took John's side.

"Where is Hassan?" Pete asked.

"Checking his plants," John answered.

Shane walked back. John looked at him and yelled, "What is he doing here?"

"I have to go," Shane rushed.

"Why the rush?" John blocked the way.

"What is going on?" Andrew looked confused.

"Why don't you wait for Hassan to tell you?" John added

"I am leaving," Shane replied.

"You are not going anywhere until we are sure you haven't taken anything," John gave a stare.

"Shane, what is going on?" Andrew turned to Shane.

"Shane! That is new. His real name is Marco," John said.

"Someone tells me what is going on?" Andrew screamed.

"He is another Patrick," Hassan replied while coming in.

"I don't understand." Andrew looked upset.

"Marco, you need to leave. I ever see you here, you know what I am capable of. Don't mess with my family. Do you understand?" Hassan warned Marco.

Marco left the house, and Andrew was shocked. He didn't understand what is going on.

"Someone tells me what is going on?" Andrew had tears in his eyes.

"His name is Marco. He was using you emotionally to get what he wants. You deserve better," Hassan patted Andrew.

"Am I stupid enough to fall for the wrong guy again?" Andrew cried.

"You are not stupid. Marco knows his game and creates this perfect illusion of the ideal guy for everyone he met," Hassan hugged Andrew and moved his hand as a sign for everyone to come and join for a group hug. Andrew wiped his tears and said, "I love you all."

Then, he looked at Hassan and said, "Tell me, how was your trip?".

"It was great. I lived all my fantasies," John said excitedly.

"So, you had sex on the beach. What else. Spill the beans," Andrew was intrigued.

"I don't believe you. You told everyone!" Hassan blushed.

"Yes, I did. Guys, you should see this video," John took out the camera.

"Stop it," Hassan put the camera down.

"No way, I want to see," Andrew grabbed the camera.

Everyone watched the video, and Hassan felt embarrassed. He buried his face on John's chest. Everyone was laughing. John looked at Hassan's face and said, "I love you.".

"So, what happened with you guys?" Hassan asked.

"I have a boyfriend now," Ethan yelled excitedly.

"Congratulations. I knew it," Hassan hugged Ethan.

"Kyle, any news?" Hassan turned to Kyle.

"I am happy," Kyle smiled.

"We talked, and everything is great between us. I believe we are meant for each other," Cara added.

"That's great," John smiled.

Everyone looked at Andrew.

"I am here sitting between you guys. I thought he liked me," Andrew looked upset.

"Look at you who doesn't like you. You are funny, smart, cute, and sexy," John responded.

"You forgot to mention stupid," Andrew added.

"Don't beat yourself because of someone not even worth it," Hassan said.

"I don't want to end up alone again," Andrew sighed.

"You are not alone," Pete put his hand on Andrew's shoulder.

"There is someone in my company who wants to know you, but I thought you are not interested," Kyle said.

"Now, you are telling me," Andrew blurted.

Kyle promised to give him Andrew's number. Andrew smiled, and everything was back to normal.

"I need to sleep. Tomorrow we have an appointment with Dr. Mario," Hassan said.

"Can we reschedule?" John asked.

"I will go and will inform you. Is that okay?" Hassan suggested.

"Sounds great," John said.

Hassan went to his room to get some rest. John made sure he was in the room so that he won't hear his plans.

"I need your help all," John asked.

"He wants to propose tomorrow," Andrew whispered.

"Why are we whispering?" Pete whispered.

"He doesn't want Hassan to hear his plans," Ethan stated.

"Stop whispering. It is creepy," Kyle said.

Everyone laughed.

"Tomorrow, I will propose to him at a romantic diner here at home," John informed them about his plans.

"What do you have in mind?" Andrew asked.

"I have many ideas, but I am not sure exactly," John looked confused.

"Leave it for us. Just focus on the proposal," Cara suggested.

"I am very excited for you. "What do you have in mind?" Kyle asked.

"I want something simple but very romantic," John stated.

"Aye, aye, Captain!" Cara responded

"You kids make me proud," Andrew looked happy.

Everyone was looking at Andrew, "Welcome back!"

Everyone laughed and discussed what they have to do and how John will pop his question. Everyone looked excited.

In the morning, everyone was happy and enjoying their breakfast. Hassan was anxious, but he didn't show. He looked at John and smiled. He then left home to the restaurant. But instead of going to work, he was on his way to Lake Ontario Park. Hassan couldn't figure out what is happening to him, why he was anxious. Everything seemed in slow motion. Finally, he arrived at Lake Ontario Park and sat on the bench.

"You don't look well," Nie said.

"He is going to propose today. I know that," Hassan said.

"I won't say congratulations," Nie said.

"Why? Don't you want me to be happy?" Hassan asked.

"More than anything in this world," Nie had tears in his eyes.

"Then, why? I love him, and he loves me. That is enough for me. It is time for you to move on," Hassan declared.

"No, you're mistaken. It is not the time. Did you ask yourself why you are feeling anxious? You already know the answer, but you are afraid to admit it," Nie said.

"I can't lose him," Hassan responded.

"You can, and you will," Nie said.

"Why don't you let me go?" Hassan asked.

"I will move on after you find the right guy," Nie declared.

"John is the right guy for me, and this is my wedding plans, Hassan took out the album and continued, "I don't care what you or anyone thinks. I know he is the one."

Hassan held the album tight and looked at the water. He spent hours sitting there and skipped the restaurant. Finally, he stood up, walked out of the park, and took an Uber to his appointment. He arrived at the medical center. He waited for fifteen minutes; then, the assistance informed him that Dr. Mario is ready for him.

"Hi Hassan, I hope you are doing well," Dr. Mario greeted.

"I am doing well. How are you?" Hassan responded.

"I am doing fine. Where is John?" Dr. Mario asked.

"He couldn't come. He has some other stuff to do, but I am here," Hassan answered.

Dr. Mario started talking to Hassan and explained to him. Hassan was listening and bleeding inside.

He was listening carefully, but it was painful. Hassan was looking at Dr. Mario closely, thinking, "Is he serious?" Hassan was scared, terrified, and feeling immense pain. The pain wasn't in one spot in his body but was everywhere. Hassan felt the world stopped for a moment. At this moment, he moved from his chair and looked at himself sitting at the chair with that sad look on his face. He walked to the window and walked back to the chair and the world back to normal. Hassan took the results in two envelopes with the prescriptions. He walked out of the room and kept walking. He cannot remember how long he walked. He felt like the world was moving very fast, and he was moving very slow. He felt pain in my chest, severe pain, and it won't stop.

Nothing is more frightening than a fear that he cannot name. He held himself together and walked to the subway station, then get a bus. Everyone who saw him on that day knew there was something very wrong with him. All in his mind was one question, "Am I damned in this life?" He got off the bus and walked back to Lake Ontario Park. He was looking at the sunset and took a deep breath. After a while, he got up, opened his phone, and ordered an Uber because he felt everyone was looking at him, pointing at him. He got in the car, and his mind was crowded with his thoughts, but one question he had to answer, "What am I goanna do now?"

Chapter 12: Threshold

Letting go & moving on!

"The truth is, unless you let go, unless you forgive yourself, unless you forgive the situation, unless you realize that the situation is over, you cannot move forward." — Steve Maraboli

Hassan reached home and stood in front of the main door but he couldn't enter. His legs were numb. Rain drops started pouring from clouds. Few drops fell on his face too. He closed his eyes and raised his face to the sky. Every drop added more to his pain. He stood there, in front of the door, thinking, "This door is the beginning and the end of my pain. But I have to deal with the pain first."

John was behind the door. He didn't know that Hassan was about to crush his dreams; the day he had waited for, for so long, would be the beginning of the end of their relationship. Everyone was happy and excited, but they had no idea what was about to happen. One side of the door was full of happiness and joy, and the other side was full of pain. All John was thinking was, "This is the moment I've waited for, for so long." Hassan put his hand on the door and he could feel the joy from the other side. The world was spinning around the door. Joy and pain, two different worlds, collided and crashed on that door. It was now pouring with heavy winds. The temperature had suddenly turned cold, like a storm of rage and pain. The storm outside was similar to the one brewing inside Hassan's heart, cold and windy. He turned and put both his hands on the door and closed his eyes. He felt the adrenaline rush, the rage, and the sorrow. Finally, he took a deep breath and opened his eyes. Now, the only thing left to do was to open that door. He took his key out, opened the door, and walked inside. He saw John wearing a dark tuxedo holding a red rose. He stopped right there, and John walked to him. John gave Hassan the rose and took his bag, and put it on the table. There was music playing in the background. John took Hassan's hand and began dancing. Hassan looked at John's face like he was memorizing it. They swayed and turned round and round. Hassan held John tight to warm his heart, enlighten it before he could crush John into pieces. Hassan wished the song could last forever, but not all wishes came true as the song eventually came to an end. Hassan took a deep breath. John got down on one knee and asked the question that Hassan had been dreading, "Will you marry me?"

In a moment, thousands of thoughts crowded in Hassan's mind, but there was only one answer for John's question, "No."

Everyone was shocked. The answer shattered John's heart into millions of pieces. He stood up and asked Hassan in a weak, sad voice, "No, is that a joke?" Hassan replied, "No, I won't marry you."

This time, each of those million pieces were crushed into further million other pieces. John was standing confused, unable to understand what was going on. His tears began to fall but in silence. Then he said, "I knew you would freak out, but do I dare to ask, why?"

Hassan was crying too, "Did you sleep with other people?"

"What?" John asked.

"Did you or did you not sleep with other people?" Hassan raised his voice.

"Yes, I am not proud of what I did. But it all happened because of the pictures I received," John confessed.

"What pictures?" Hassan asked.

"I cannot tell you," John said.

Hassan stepped back, "Is it even real?"

"He is right. I saw the pictures. Don't blame him," Andrew backed John.

Hassan looked at everyone and said, "You all need to leave us alone."

"No, I want them to stay," John interrupted.

"I asked them to be here for your sake, but why not stay and watch me crush him. Anyway, where are the pictures?" Hassan asked again.

John nodded in disagreement. Hassan couldn't control it and yelled, "Where are the damn pictures?"

John switched on his phone and gave it to Hassan. Hassan couldn't believe his eyes, and tears started falling from his eyes. He put his hand on his mouth and cried aloud. Hassan raised his head and said, "Why didn't you tell me?"

"I thought you were cheating on me," John answered.

"Really! Are you that dumb? Someone took advantage of me while I was under medication, and you wanted to hurt me back by sleeping around," Hassan asked.

"I don't know what to say," John was clueless. "I am sorry," John continued.

"Me too," Hassan said.

"You promised me," John lost his temper and yelled.

"I promised 'John', the person I loved. I don't even recognize you anymore," Hassan responded.

"But I love you," this time, there was a softness in John's voice.

"I love you enough to let you go, and you should too," Hassan replied.

"We can fix us," John held Hassan's hands.

"So, you want to fix us?" Hassan walked to the table and opened his bag, took two envelopes out, handed them to John, and said, "Tell me, how you can fix this."

John took the envelopes and opened one after the other, and read it carefully. He broke down and fell to his knees, "No, No, No."

Hassan looked at him and said, "Congratulations, your plan worked. You hurt me and destroyed us." Hassan looked at everyone and said, "The show is over."

Hassan sat on the floor with his back to the wall and shut his eyes. Everyone ran to read what was in the envelopes. They saw the reports of both of them being HIV positive. Pete moved to Hassan and hugged him. Hassan looked at everyone and said, "You don't need to pick a side. He was a victim, so am I."

Hassan stood up and took his album from his bag, and went into the kitchen to burn it on the stove. He watched it burn. He wished he could burn everything to ash so he could forget this pain. John looked at Hassan burning the album and felt an immeasurable pain. He walked out of the house and drove to his mom's.

Hassan entered the living room and started breaking things like crazy. Cara tried to stop him, but Andrew stopped her from interrupting, "Leave him be. He is angry because he lost his love."

Everyone was watching him crying. After a while, he stopped and looked around; he saw John with him in every corner of the house. He fell on his knees, and Andrew ran to him and hugged him.

The storm outside hadn't stopped, but the one inside Hassan's heart had come to a standstill as he lost conscious in Andrew's arms. Everyone rushed towards Hassan with tears in their eyes. Andrew quickly took him to his room.

Hassan opened his eyes and found himself surrounded by all his loved ones except John.

"Are you feeling better?" Cara asked.

"How can I feel better?" Hassan responded.

"No one saw that coming," Kyle said.

They all showed concern, and Andrew asked Hassan to start taking his meds.

"Are you angry because John gave you HIV?" Andrew asked.

"No, I am angry because he hid it from me and lied to me. I am angry because I lost him," Hassan was upset.

Hassan looked up and said, "I need to be alone."

"I won't leave you, but they can," Andrew persuaded.

"Don't worry, I won't do something stupid like last time," Hassan promised.

"Then let me stay with you. They can clean the mess you made," Andrew asked.

"I am sorry, guys," Hassan apologized.

"Don't worry," Cara smiled.

They all left the room, except Andrew. He looked at Hassan and said, "Now you can cry." And he did. Andrew held him in his arms; he could feel him shaking. After a while, Hassan stopped crying. He closed his eyes and opened them only to find himself sinking in the water, unable to breathe. He closed his eyes again and saw John crying. He opened his eyes again and found himself in his room. He stood up and went downstairs, and Andrew followed him. He searched for his phone in his bag to call John, but he didn't answer. Hassan threw his phone, and the sound grabbed everyone's attention. He then raised his head and started crying in silence. It was a heartbreaking scene to witness.

Andrew walked up to him and hugged him again. He comforted him and asked him to try and sleep and that they'll talk later.

***

When the sky is totally covered by the dark clouds, be strong enough to see the bright stars beyond them! — Mehmet Murat Ildan

Nine days had passed, and the dark clouds still surrounded the house. No one had said a word that morning until Hassan received a call from Marguerite. Hassan looked at the phone and thought he was dreaming until he answered, "Good Morning, Marguerite."

"Good morning, darling. How do you feel today?" Marguerite asked.

"I am feeling better. How are you? How is John doing?" Hassan inquired.

"I am doing okay, but John is not well. I know you don't want to see him or talk to him, but you are the only one who can help him. I never asked you anything, but..." Marguerite cried, "now I am asking you to bring my son back again. I am asking you to give me my son back."

Hassan asked about John's whereabouts, and she sent him the address. Hassan cut the call and headed to the door.

"Where are you going?" Kyle asked.

Hassan walked in silence. Cara blocked his way, "No way, we won't let you go alone."

"I will come with you," Pete said.

"If anyone of you comes, the situation will be worse," Hassan said.

"At least let me drive you there," Andrew proposed, and they both left.

Andrew took Hassan in his car and drove to the address that Marguerite had sent. Hassan didn't make a single sound all along the way. He was thinking about what he would say to John. It wasn't easy for him to talk to John after what had happened. Finally, they arrived, and Andrew dropped Hassan while he stayed in the car. Hassan searched for John until he found him sitting in a corner. He was about to inject himself with a drug. Hassan ran toward John with tears in his eyes and stopped him. John felt a weird chill. Like Hassan had touched him, he could sense his presence. He closed his eyes, and Hassan said with a weak crying voice, "Don't do that to yourself, please."

He opened his eyes and realized that Hassan was really there. He stood up and turned around, crying, "You don't care about me. You don't want to see me. Why are you here?"

"I will always care about you," Hassan confessed.

"How? I saw you burning our photos," John declared.

"You idiot, I didn't burn our photos. I burned my wedding plans along with my dreams," Hassan cleared the misunderstanding.

"Do you hate me?" John asked.

"How can I hate you? How can I hate the incredible years of my life that I spent with you? Come home!" Hassan pleaded.

"I can't go back there after what happened—they are all on your side, no one is on mine," John said.

"You are wrong. There are no sides," Hassan told the truth.

"I will come back, eventually," John looked down.

"You are coming with me now, and call your mother," Hassan directed.

John spotted Andrew, "Okay. I see you brought company."

"He drove me here, and I asked him to leave. I guess he didn't," Hassan stated.

"I quit my job," John looked upset.

"Allen called me a few days ago. So, you still have your job," Hassan smiled.

"I hurt you. How can you forgive me?" John looked him in his eyes.

"Make sure you don't make the same mistake with the next guy," Hassan stated.

"I don't think there will be another guy," John stated.

"There will be," Hassan said.

"How? I am living with HIV," John replied.

"I am living with HIV too. It is not a crime, and you can have a normal life," Hassan mentioned.

Hassan offered his hand to John, and John took Hassan's hand and walked to the car. They got in the car. Andrew looked at John, smiled, then said, "Welcome back." Andrew drove home, but they made a stop at the pharmacy to pick up John's meds. They arrived home, and everyone greeted John. They gave him a big hug. Hassan took him to his room, "I am glad you are back," Hassan said.

John hugged Hassan, and Hassan cried. John looked at Hassan's face, wiped his tears, and said, "I didn't kiss you that day. If I knew it was our last day together, I would have kissed you." Hassan closed his eyes and felt the warmth of John's hands on his face. John was now crying and they both kissed. It was nothing more than a goodbye kiss. Tear after a tear was signalling the end of a great love story, an ending that was full of pain and sorrow. It was time for the dark clouds to leave. The job was done to divide one heart into two broken hearts.

Hassan stopped John and left the room feeling tremendous pain in his heart.

He went downstairs, and everyone was staring at him. Then, finally, Andrew came to him and hugged him and asked him, "How are you feeling?"

"He is here, and I can't be with him. It's unbearable," Hassan said.

Andrew took him to the kitchen. Everyone was sitting at the table, waiting for Hassan and John.

"Where is John?" Cara asked.

"Still upstairs," Hassan mentioned.

Ethan went upstairs, knocked on John's room door.

"Come in," John said.

"You need to come downstairs and join us for lunch," Ethan asked.

"Thank you. I will be there in a few minutes," John said.

Ethan turned back, and John said, "Ethan, I am sorry I hurt you countless times. I shouldn't have taken him from you. None of this would have happened if I hadn't stolen him from you."

"Don't beat yourself up. You gave him things I wouldn't have been able to. You made him happy. Now you need to move on. You both deserve to be happy. Don't keep us waiting," Ethan patted his back.

"I will come with you," John stood up.

They both went down and joined the rest of the group for lunch. John sat next to Hassan, and Hassan gave John a key chain with keys, "This is the new key. When you open the door, you need to enter the passcode, which you will find on the tag."

"Passcode!" John looked surprised.

"It is annoying, but better," Andrew stated.

After lunch, Hassan went outside with Andrew. The others stayed at home. They were all sitting in the living room yet not a single word had been spoken for more than an hour. Then Pete decided to play a song and started to dance. He took John's hand and said, "I know you are not in the mood, but you can dance it out."

John and the others joined Pete for the dance. As soon as John started to move his body and felt the rhythm, he smiled. He closed his eyes and imagined himself dancing on the clouds. He kept moving his body and forgot about his pain for a while. Suddenly he stopped, and tears started to drop from his eyes. Everyone stopped and looked at him. They knew it was time for him to let out his pain. "I ruined everything. It's all my fault."

At that moment, Hassan entered with a box in his hands and said, "Stop beating yourself. It's not your fault. Never was, never will be."

Hassan walked to John, gave him the box, and said, "This is for you."

"What is this?" John asked.

"Open it," Hassan looked at the box.

John felt a movement inside the box. He put the box on the floor and opened it. There was a puppy inside, and John took it out and said, "C'mere, buddy." He smiled and looked at the puppy, "What's your name, buddy?"

"You need to name him," Hassan smiled.

"That's very nice of you. But why is he allowed to have a puppy, and I'm not?" Ethan complained.

"The rules have changed," Hassan winked.

"So, can I have a puppy now?" Ethan looked excited.

"I thought you would ask for a cat. Anyway, one puppy would be enough for this house," Hassan teased.

"Thank you," John said.

"You are a father now. You are responsible for your child. So now, name him," Hassan asked.

"Archie. His name is Archie," John said. Everybody welcomed Archie and looked happy, "We love you, Archie."

Everyone was playing with Archie, but he liked John the most. Hassan sat on the couch, watching John and the others playing. Ethan reminisced the past when Ethan and John used to play with John's dog, Archie. He saw the same look on John's face now. He then turned to Hassan and wondered, "How is he holding on. It must be hard for him to give up his love." Andrew looked at Hassan and felt his pain. He felt like he was crying inside. Without wasting any time, Andrew walked to Hassan and asked him to talk in private. Hassan went with Andrew to talk.

Without a single word, Andrew hugged Hassan and said, "It's alright. You can let it go now." Hassan started crying. Andrew wiped Hassan's tears and got closed to his face, and kissed him. Hassan didn't move, and Andrew kissed him again. He was holding Hassan's face between his hands. He looked at Hassan and said, "I always wondered what it would be like to kiss you." Then continued, "I was right. I've kissed many people in my life, and no one holds a candle for you. You are so pure and raw. Kissing you is like kissing someone with a soul. And it tastes delicious. You are very different. Now I understand why John was crazy about you."

Then he reached Hassan's lips and kissed him again. It was a long kiss. Hassan didn't say a word or move. Andrew wrapped his arms around Hassan and said, "Can I tell you a secret?"

Hassan nodded.

"I was jealous of you because I thought you had it all, but I was wrong," Andrew said.

"I lost him," Hassan sighed.

"You will find someone who will love you," Andrew assured.

"I don't think so. Who wants to be with someone like me?," Hassan looked upset.

"If you take your meds responsibly, why not? You are sexy, smart and a very good person. Who doesn't want to be with you?" Andrew questioned.

"Not all people think like you," Hassan said.

"Let us make a deal if you didn't find someone, we get together," Andrew proposed.

"You think so?" Hassan asked.

"Let's seal our deal with a kiss," Andrew winked.

"You are crazy," Hassan smiled back and continued, "But I don't think it is a good idea."

"Just shut up and kiss me already," Andrew teased.

"Woah, you've got a temper," Hassan smiled, got close to Andrew's lips, and they kissed. Then, Andrew stopped and said, "Now I know why John calls you shorty."

"Can I tell you something?" Hassan asked.

"Sure, anything," Andrew said.

"You are the third guy I've ever kissed," Hassan smiled with tears in his eyes.

"You are kidding!" Andrew took a step back.

Andrew looked at Hassan and realized he was telling the truth. He smiled and said, "I hope there won't be a fourth one."

"You are mean," Hassan laughed.

"So, you will be mine. By any chance, will you get any taller," Andrew teased him.

"Dark and mean," Hassan gave him a stare.

"I know," Andrew laughed.

Andrew and Hassan went downstairs and played with Archie for a while. Ethan was looking at Andrew, and many thoughts crossed his mind. He took Andrew aside to talk, "I thought you were a good guy." Ethan said.

"What are you talking about?" Andrew looked confused.

"He is vulnerable, and you kissed him. What were you thinking?" Ethan asked.

"I hate to break it to you, but I have no intention of hurting him. Has it ever occurred to you that I might like him?" Andrew asked.

"He just broke up with John and he's not ready to move on," Ethan said.

"You are right. I just want you to know I am not a bad guy," Andrew responded.

"Act like it," Ethan replied.

"Okay, then. You should know we had a deal. In the future, if he doesn't find anyone, I will be his boyfriend," Andrew talked about the deal.

"What are you talking about? How is that even possible? He is a sexy, smart, and caring person who doesn't want to be with him," Ethan asked.

"That's what I told him, but he's stuck with the idea that it's the stigma of the virus that makes him undesirable," Andrew sighed.

"Yeah, that is a problem. But unfortunately, people always will judge harshly," Ethan accepted.

"He will find someone who will appreciate him," Andrew looked confident.

"Let's get back to them," Ethan smiled.

Andrew and Ethan joined the others. John was happy to be a dad for Archie, and Archie looked happy too. He was playing, jumping, following John, and this made Hassan smile. Cara looked at Hassan, sat down next to him, hugged him then kissed him on the forehead. John looked at Hassan with a smile on his face and whispered, "Thank you."

Hassan smiled and said, "I feel tired. I will go take a nap." Hassan went to his room, and Ethan followed him. He found him crying in silence. Hassan wiped his tears and said, "I am sorry. Do you need anything?"

Ethan sat next to Hassan, held him, and said, "I wanted to check on you. Are you alright?"

"He is here, and that's is important. Everything else doesn't matter," Hassan fake smiled.

"You're mistaken. Your feelings matter. I know you want to run to him. I know the pain could be unbearable, but you are Hassan. You are the strongest person I've ever met," Ethan held his hands.

Andrew entered and said, "Yes, you are."

"I will let you both talk," Ethan said and left the room.

Andrew sat next to Hassan and hugged him, "Would you mind if I take a nap with you?"

Hassan nodded his head in agreement. Hassan fell asleep in Andrew's arms. John entered the room to check on Hassan and found Hassan sleeping in Andrew's arms. At that moment, John wished he could turn back the time and do everything differently.

"I think he's used to sleep between your arms?" Andrew presumed.

"Yes," John sighed.

"Since you both broke up, he has trouble sleeping," Andrew told John.

"Please take care of him," John requested.

"Why do you want me to take care of him?" Andrew asked.

"You are the only one who can understand him completely," John asserted.

"I will do my best," Andrew assured.

John left the room to go to his room. He didn't even close the door, as he ran to his bed and cried. Cara, Pete, Kyle, Ethan, and Archie followed behind him. Pete put Archie on John's bed, and Archie started licking John's face. He looked at Archie, held him between his arms, and said, "Daddy is a little sad, but he is happy to have you. Let's go for a walk." John took Archie for a walk. John's feelings were twisted between happy and sad. After the breakup, his life had taken an unexpected new turn. No one has ever had that effect on John's life like Hassan, but it was all over now. All those moments were now turned into memories. A candle burned till the end, leaving a lovely, unforgettable memory behind, and Hassan wasn't an ordinary candle. This candle left a memorable moment behind, along with the touches and fragrances.

Hassan was sleeping in Andrew's arms. He had a dream where he saw a handsome man dancing with him and telling him that they would be together one day. The man didn't tell him when but told him everything happens at the right time. Just when the music stopped, the man left, and at this moment, Hassan woke up. Andrew looked at him and asked, "Were you dreaming?"

"Yes, it was weird but nice," Hassan smiled.

"You can tell me about it," Andrew asked.

Hassan started telling Andrew about what he saw in his dream. Andrew was listening. He smiled and kissed Hassan on the forehead, "So, do you feel better?"

"Yes, I do," Hassan looked happy.

Chapter 13: Kill the Silence

"For many survivors, breaking the silence is an important part of the healing process"
- Melissa Kleinman

Several months had passed, and Ethan's birthday was now coming up. He met up with Pete at work, and they both planned a party for his special day. While Pete shared his encouragement over Ethan being excited about the party, he couldn't help but wonder that it's very unlike him to care about such events. He asked Ethan if it's because of someone he's hoping to meet there, but Ethan confided in him that he's actually hoping to break John and Hassan's loneliness and has invited many of his colleagues and people from another department as well so they both have someone to chat and hang out with. Just as they're both done going over their plans for the party, Camille from HR walked into the room. Pete is dumbfounded as to what she's doing here. She asked him if it's true that Ethan's throwing a party for his birthday. Pete didn't know what to say, so he told her to ask the man himself and excused himself frantically out of the room.

"Yes, Camille, you heard right, there's going to be a party. And it's at Hassan's place."

"Oh, I didn't know you guys were living together," said Camille.

"Yeah, I live there with my boyfriend," Ethan tried to clear her doubts.

"Can I ask you something, but please keep it between you and me."

"Sure."

"Do you know if Hassan is seeing someone?"

"No, he is not seeing anyone. He is single."

"That's good then. Because I wanted to introduce my brother to him. You know what I mean."

"Great, no problem. Bring him to the party." Ethan told her assuredly.

"One more thing," she stopped herself on her way out to ask Ethan another question.

"What is it?"

"Do you know if Hassan would have a problem with my brother?"

"Why would he have a problem?"

"It's just his skin color. I'm worried he might have a problem with it."

"Oh, you don't have to worry. It's not a problem. Hassan is not that kind of person."

"Thanks, I just wanted to make sure."

"Just bring him."

"Thank you." Ethan's hopeful words seemed to have lifted the stress from her mind.

As soon as she left the room, Pete came back inside and asked Ethan why did he invite her brother, let alone herself, to the party, to which Ethan responded that she's in the management team and it would have been unwise to say no to her. "Besides, what's the worst that could happen?" he asked Pete.

Meanwhile, Brian was standing outside the room and was eavesdropping on their entire conversation. He felt an incredible rage build up inside him and quickly ran over to the restroom to calm himself down. There, he rinsed his face a couple of times, then looked at his angry reflection in the mirror. The only thought that lingered inside his mind was believing that Hassan belonged to him and they're destined to be together.

At lunchtime, Brian went to talk to Hassan to try to smother him. He sat next to him and tried to strike up a conversation.

"Can I ask you a personal question?"

"Really, here? We're in the lunchroom. Do you think it's the right place to talk about personal matters?"

"Okay. Then will you be available after work? I will take you out for a drink."

"Yeah, sure, why not."

"See you then."

"See you."

Brian moved away disappointedly and Hassan went back to his desk. Ethan came in and asked him to come to his office for a moment, and Hassan followed.

"Pete wants to throw a birthday party at your place. You know about this?" Ethan told him.

"Why does everyone want to talk about private matters here at work? We have a house. We can talk about it there," responded Hassan.

"Relax! I just wanted to give you a hint."

"Alright, well, thanks for the hint. Can I go back to work now?"

"Sure, Mr. Hissy Fit."

Hassan returned to his work. Right this moment, Tristan came over to Hassan's desk to talk to him about Ethan's birthday party. "Did you hear that Ethan is throwing a birthday party?" he asked him, much to Hassan's annoyance. "Yes, I do," Hassan responded in a condescending tone. He just couldn't seem to catch a break from all the gossiping. "Are you invited?" inquired Tristan, to which Hassan replied, "Why wouldn't I be? the party is at my place." Hearing these words shocked Tristan as well, he also tried to confirm his suspicion, "Oh, are you two living together?" Hassan began to grow impatient and raised his voice at him slightly, "Why are people asking me so many personal questions today? Anyway, he is my roommate, and he has a boyfriend of his own."

"Great." Tristan began to leave the room.

"Am I missing something?" asked Hassan curiously.

"Nothing. You don't seem to be in a good mood. It's just that everyone was talking about why did you break up with your boyfriend," said Tristan, jumping right back into the conversation.

"Oh, Tristan, you just love to gossip, don't you?" said Hassan annoyed.

"No, I don't. Jeez, I'm only asking you as a friend." Tristan defended himself.

"You know we are at work," Hassan felt the need to remind him.

"Come on, give me some details. "He persisted, further aggravating Hassan.

"Alright, we had a fight. Is that enough for you?" said Hassan finally.

"I don't think that is the reason. "It seemed like Tristan was hell-bent on finding out the details.

"My personal life …. is not anyone's business. Am I clear?" This time, Hassan said sternly.

"Crystal. I am sorry," said Tristan as he felt a little embarrassed now at his interrogation.

"Don't be. I am just tired of this question, and when I am tired, I am mean," Said Hassan, trying to make up for his tone.

"Do you know why people talk about you so much?" 'This guy clearly can't take a hint,' thought Hassan, then asked him in a sarcastic tone.

"I don't know. Why don't you enlighten me?"

"Well, for one, many of them are jealous of you, while others hold an interest in you."

"And which one are you?" he asked Tristan.

"I like you, but I didn't make a move because I don't know if you are interested in someone like me."

"Someone like you?!" Hassan asked him with shocked expressions on his face.

"Isn't it obvious?" Tristan sighed.

"Ethan's boyfriend is like you, and honestly, I see you as a fine man. But you need to stop gossiping around so much." Hassan advised him.

"If I asked you out, will you go with me?" Tristan was aiming for something.

"It's not the time, nor the place, but I would like to go out with you, but I am not ready yet to start dating." Explained Hassan

"I understand." Tristan nodded his head in agreement.

"Thank you," Said Hassan, hoping that Tristan will now leave him alone.

Tristan finally returned to his desk, and Hassan took a long sigh, then focused on his work. Fortunately, he was not bothered by anyone for the rest of his shift. After work, Brian came to

Hassan, and they both left together. Brian took Hassan to a park to talk. There, he looked at Hassan's eyes and smiled.

"You know I like your eyes."

"Thank you, Brian. Now, what did you want to tell me?"

"Did you start dating?"

"No, I don't think if I will ever be ready."

"Why?"

"Can I tell you a secret?" asked Hassan in a highly emotional tone.

"Sure."

"I am living with HIV," Hassan said while weeping.

Brian looked at Hassan and hugged him. "Why did you keep this a secret from me?"

"You've been very supportive of me ever since John and I broke up. But I am still not ready to move on." Hassan confessed to Brian while wiping a tear that was running down his cheek.

"I know. But I want you to remember that I will always be on your side," Brian patted him on the back.

"Thank you, Brian." Hassan seemed to be in a much better mood now.

Afterward, Brian and Hassan continued their walk in the park, and then Brian drove Hassan home. At his own house, Brian went to the basement, where he had a painting of Hassan placed next to a big mirror. He hugged the painting amorously and then began running his fingers on the canvas while saying, "I know you are screaming inside like me. I know you belong with me."

***

Kyle had just reached home after buying flowers for Cara when he received a call from his ex-girlfriend, Lily. Kyle answered the phone. "Hey, Lily."

"Hey, Kyle. How are you doing?" Lily asked him.

"I am doing great, thanks. And yourself?"

"I am doing okay. I am just wondering if you have time to meet and talk."

"I don't know if I have time to meet you."

"Oh, okay."

"Wait, Kyle, can we meet at the same place at 9:30?"

"Sure, I will be there," Kyle said with a hopeful smile.

"Great. See you then."

"See you."

Kyle hung up the phone, then entered the house and surprised Cara with the flowers. Cara was happy to see the beautiful flowers; she smiled and kissed Kyle to thank him.

"I just received a call from Lily," said Kyle.

"Lily? Who is Lily?" asked Cara curiously.

"She's my ex-girlfriend."

"Oh!" Cara's smile turned into a frown.

"Oh, honey, please don't worry about anything. Besides, I want you to meet her with me."

"Why?"

"I don't know what she wants to talk about, but I want you to be there. I won't hold any secrets from you."

"Okay, I will come with you," Cara said with a smile. Her worrying seemed to have lessened by a lot now.

"I love you," Kyle said passionately.

"I love you too."

Hassan entered the room while Kyle and Cara were sharing a passionate kiss.

"Hey, you two love birds. We are all waiting on you."

"We are coming." Cara and Kyle said in unison.

The couple joined the others at the dinner table.

"Any interesting stories from today?" Kyle to everyone sitting around. Hassan said he'll start and told everyone how he was continuously disturbed at work today, "Today, at work, I was not able to catch a break. Everyone kept hankering me and asking me questions about Ethan's birthday party."

"Since when are you into birthday parties?" Kyle asked Ethan.

"Relax. It's not a big deal." Ethan replied to him. Pete chimed in and tried to defend Ethan. "I want to celebrate my boyfriend's birthday. Is that so wrong? "Hassan jumped in again and said, "It's your right to celebrate your boyfriend's birthday. But I don't understand why your boyfriend had to invite half of the company. "Cara tried to clear Hassan's confusion and told him, "Okay, I will tell you but promise me you won't be angry."

"That depends. Do I need to be angry?" Hassan asked her condescendingly. Seeing Hassan's judgmental expressions, Kyle thought it best to answer on behalf of Cara, "Well, to be fair. Since you and John broke up, you both haven't had a chance to meet anyone." Hassan stopped eating and then looked at his plate in contemplation. He understood and even sympathized with Cara and Kyle's plan. John looked at Hassan and said, "I think they are right. I don't want you to be alone." Hassan looked at John and responded negligibly, "They are right, but I am not ready to start dating." Seeing the awkward situation developing, Pete told Hassan, "At some point, you need to be open to meet others."

"I know, but I am not ready. I am not there yet." Hassan responded while still looking down at his plate and eating slowly. "At least you can try," Ethan said to Hassan confidently. "Alright. But don't blame me if I end up ruining your party."

"Listen. All I want is to see you both moved on and happy," said Ethan as John got up from the table. Hassan stopped him and asked him, "Where are you going?"

"I just can't. Not right now," said John remorsefully. "Sit down and finish your dinner. After dinner, we can talk," Hassan tried to calm him down. Seeing the tension developing, Andrew shared what's been on his mind, "I was silent the whole time, but now I will speak. Look at you both; we don't recognize any of you. Hassan, where is the smile we used to see on you? John, stop blaming yourself and move on. What happened to you both? Why are you avoiding other people? At some point, you need to realize that you both are adults and responsible for your actions. Just shut up and start dating already. "Hassan understood where Andrew was coming from but still remained defensive, "I see your point, but has it occurred to you that I might just not be ready to jump back in the dating game again?" Andrew couldn't stand Hassan's reply and said to him, "Don't give me that excuse. You are both ready. Hassan, your stubbornness won't help you to get over John. John, stop dreaming that you could back together. Wake up and embrace what you have, and move on. Life is short to waste time." No one said a word for the rest of the dinner, and everyone tried to finish their meals quickly to escape the situation. Nevertheless, it was written on every person's face at the table how awkwardly they were all feeling.

After dinner, John took Archie for a walk. Cara and Kyle went together to see Kyle's ex-girlfriend. Ethan was helping Pete in washing the dishes, and Hassan was listening to Andrew playing the piano. Andrew asked Hassan to sit next to him, so he got up and moved next to Andrew. Hassan liked what Andrew was playing. He took a deep breath and closed his eyes. He felt the rhythm and danced freely to the tune in his mind. Andrew finished and wrapped his hand around Hassan. Hassan opened his eyes and said, "That was nice." Andrew took Hassan to his room and closed the door. Hassan looked at Andrew, wondering what was happening. Andrew sat next to Hassan and asked him, "Be honest with me. Do you want to end up with me?"

"I don't know, Andrew. I still believe there's someone out there for me."

"Hassan, stop acting like an idiot. Just open your heart and live your life. Allow yourself to be happy for once." Andrew told him disappointedly.

"I am trying," Hassan responded. Andrew gave Hassan a long, determined look before moving closer to him,

"What are you doing?"

"I'm trying to kiss you. Can I?"

"No, Andrew, please don't."

"Come on, open your heart. I might be the one." Hassan got close to Andrew's face and said, "I am too much for you," then stood up and began to leave. Andrew held Hassan's hand and tried to drag him towards the bed. "You can't stop me," he said. "And why is that?" asked Hassan. "Because you need a human touch to remind you that you are alive. Everyone needs it. So now, can I kiss you?" Andrew pleaded consistently. Hassan looked at him and smiled. Andrew bit his lip, smiled, and said, "You are really too much for me. I can't handle you." Hassan stood up, turned to Andrew, and said, "I told you."

"I want my kiss," Andrew said before leaning forward and putting his hand around Andrew's face. They both shared a long kiss that left Andrew panting. Andrew flipped Hassan on the bed and continued kissing him. Hassan kissed him back and received a compliment from Andrew, "You are a good kisser." Before their romance could develop into something more, Hassan stopped himself, threw a mysterious look at Andrew, and went upstairs. There, he played a song on the TV and danced to it. Pete, John, Ethan, and Andrew, all joined him for the dance. Archie also joined soon after and began jumping around them. Hassan stopped to take Archie in his hands and kissed him. Then he put him on the floor and tussled around with him. Everyone was watching Hassan. John smiled and joined Hassan to play with Archie.

It's now a Saturday, and everyone was busy preparing for the party. Soon, people started gathering inside the house. John seemed incredibly anxious as if he was afraid that people might ask him about his breakup. Too many thoughts ran through his mind and continued to stress him out. But, through it all, the one thing that calmed him down was Hassan's smile, who was welcoming the guests seeming cheerful. Camille also arrived with her brother Roy and walked towards Hassan and introduced Roy to him. Brian was watching and boiling inside after he saw Hassan talking and laughing with Roy. He went to use the washroom again to calm his nerves.

"There are so many people here. You must be very popular." Roy told Hassan.

"Don't fret. Most of them are from work."

"You know, my sister told me you broke up with your boyfriend. I can't imagine why any person would want to leave such a good person like you."

"We've only just met; you don't know the type of person I am."

"True, but I can sense people's vibes very early on."

"Thank you for your kind words. Now, please excuse me." Hassan felt too claustrophobic and went to his room to clear his mind. Roy understood Hassan's reaction and realized it was too soon to talk about the breakup. Hassan was staring at the trees through the window in his room and contemplated his relationship with John. Soon after, he heard a faint voice reach him from the back, "Mind if I join you?" He turned around to see Brian standing at the door with 2 drinks in his hands. "Yeah, sure," Hassan replied while straightening his arched posture from bending down near the window. Brian came over and offered Hassan one of the drinks he was holding. "Thanks, although I have been trying to stay off of margaritas."

"I am not so sure I wouldn't even call these margaritas. Pete spares no expense with the sugar when it comes to these." Hassan chuckled in response. "So, what's got you down? Why are you not celebrating like the rest? Is it something against Ethan?"

"No, it's nothing against him. It's just…after what happened with John, I have just been finding it really hard to…just be around people."

"It's alright. Hell, I don't know whether I should be sad for you or sort of relieved that you're not with John any longer."

"What do you mean?" inquired Hassan.

"I mean, you're so beautiful and handsome and gentle, and I don't think you deserve to be with anyone else who doesn't see you that way."

"And you do?"

"Oh, yes. Your soft, creamy cheeks, your mesmerizing eyes, and your beautiful, hairy body."

"Thank you for your… words. But the last thing I want to get involved in right now is another relationship. Wait, how do you know I am hairy?" Hassan asked him curiously. He tried to remember when did he ever tell him about this fact. Brian tried to change the subject, but he knew uttering those words had pushed Hassan to the deep end. Nevertheless, Hassan remembered what Pete had told him, and after hearing what Brian said, he was able to piece together the complete story. "It was you, wasn't it? You sent the pictures. You're the reason John had been so distant from me. You're the cause of all this misunderstanding and confusion."

"Listen, I can explain." Brian tried to justify himself, but it was no use since Hassan was visibly furious at him. "I can't believe you would have the nerve to be even standing in front of me, to come here tonight. I don't want to hear a single word. Please get out of here. "Brian drew closer to Hassan and put his hand on his shoulder. "I've been in love with you ever since I've known you. I …'' But before he could speak anymore, Hassan immediately shrugged off his hand and rejected all of his advances. He opened the door and said, "Get out of this house, right now!"

Brian replied laughingly, "Do you think anyone will accept to be with you? Wake up, and don't miss your chance." Hassan was unable to stand the sight of him and responded confidently, "You are right. I have to wake up, but first, I will make you taste your own medicine. The difference between you and I, I don't hide and stab from behind. And don't be worried about me because I am Hassan."

Hassan made his way to the door. He found Cara and Andrew, "Sorry, Cara, but your brother needs to leave the house and never come back again," he told them. Hassan walked downstairs angrily, with tears in his eyes. He wiped his face clean and stood on a chair. He then spoke to everyone at the party. "Can I get everyone's attention?"

The people at the party started looking at him curiously. At first, there was some confusion, but eventually, everyone stopped talking and looked at Hassan as he addressed them.

"Ethan, I am sorry for what I am about to do. I know you will understand. Everyone, you know Brian, right?" he let out a sigh while pointing to him. "He works with us. He's noticed that some of you have an interest in me. I am really flattered, but no, I won't hide anymore. The truth is I am HIV positive. So, if you are interested in me, you need to rethink. And even if anyone can't see me as a whole person, you are not welcome in my life."

Hassan got down from the chair and said to Brian that he should leave before walking out into the yard toward the swing. When everyone turned to look at Brian, Cara asked him to go and walked with him towards the door to bid him farewell. Brian tried to make his plea and justify his actions to her.

"He is mentally…" before Brian could complete his sentence, Cara interrupted him mid-speech, "Don't even say a word about him."

"This is how you treat me? I always thought I could rely on you. You are my sister."

"I'm his sister, not yours. You can view me as whatever you want, but I will always view you as an asshole and a homewrecker for what you did here."

"You do not understand me. You're too enamored by his sultry and sheepish words."

"Just shut up, alright? Don't say a single word. Not until you realize the reality of your actions tonight, don't bother coming back to any of us ever again. Even if I could somehow forgive you for what you had done to me, I will never, ever forgive you for what you did to him."

"Cara, I know you think of me as a monster, but there was a reason behind what I did. I never meant to cause you any harm. I care for you deeply."

"Monster? No, you're not a monster. Even monsters can grow to realize the errors of their ways. You're a specter without a soul. You drew my husband away from me at a time when I needed him the most. He held me close to him, he kept me warm, kept me hoping that I'd be able to get through my pregnancy and you took that away from me. I lost my unborn child because of you, my only source of happiness, my reason to live, my sweet angel who I never even got to hold in my hands. You took my husband away the same way you tried to draw Hassan away from John. You only think you care for me, that you're there for me. But the only person who you have ever really helped is yourself. I have held and buried this anger inside of me for way too long now. But there is no reason to keep pretending I'm fine. I'm not, because of you. So, get out of this house. Get out of our lives. And don't bother looking back. There is no more refuge for you here."

Hearing her gut-punching words, Brian was unable to come up with any retort and thought of it best to wipe the tears off his face and bid farewell to Cara and everyone else at the party.

After what happened with Brian, many other people also began to leave. Cara went to her room and opened her drawer; she took out a sonogram and looked at it. Soon, Kyle entered the room and embraced her from behind. "I heard everything," Kyle said. Cara burst into tears, Kyle held her even more tightly and comforted her. After a while, when Cara had composed herself, both of them went downstairs. Pete and Ethan were worried about Hassan and what would happen next. Andrew's eyes followed Hassan around, who sat down on the swing. Tristan came over and sat next to Hassan to talk to him.

"You just told the whole company that you are living with HIV. I don't know what happened to you, but you need to keep yourself together. I don't know what to say about what you did, is it brave? Is it stupid? Or both? One thing I am sure about is, you are a strong man."

Tristan pulled Hassan into a kiss and left. Nearly all guests had left the house at this point.

Hassan looked at Ethan, "I am sorry for ruining your party. Cara and Andrew, a word, please." Hassan pulled them into his room and closed the door behind them.

"How much did you hear?" Hassan asked them.

"Everything," replied Cara.

"So, you know what your brother did?"

"Yes, and I am so sorry." Cara shared her sympathies and tried to placate him.

"Don't ever apologize, Cara; you have done nothing wrong. Instead, I want you to promise me not to tell anyone what happened here in this room."

Cara nodded and responded by saying, "I promise." Andrew chimed in, "Me too."

"Good. Could you excuse me, I need to be alone." Hassan replied. But before he could leave the room, Andrew stopped him, "But first, you need to talk to John."

"You are right," Hassan responded.

All three of them went downstairs. Hassan walked over to John, who was crying and sat next to him.

"What's wrong?" Hassan asked.

"It's all my fault. I am so sorry," replied John.

Hassan looked at John with a ferrous look and held his head between his hands. Everyone was worried about what Hassan will do. Finally, Hassan took a deep breath and said, "If I ever hear you say it was your fault or apologize for it, I will kick you from now to next year. What happened was not your fault. Never was and never will be. Do you understand me?" John nodded his head, and Hassan kissed him on the lips. "For how long are you planning to cry?" Hassan asked him.

"I can't stop," John replied through his tears.

"You know how much I love you, but you need to understand that you have to move on in your life. Our relationship now is a part of our past. We have a whole new future with others. Don't miss your chance to live your life. Whatever happened, remember, you are a wonderful person, and thousand will die to know you. And I will be the one who will walk you to your husband. I want you to be happy. Now wipe your tears and stand up. Tonight, we will go out, and we will dance. Who knows, maybe you will meet someone. You need to open your heart to new people." Hassan assuaged him with a smile before standing up and proffering his hand to John. John accepted Hassan's hand and smiled as he wiped tears from his eyes. Andrew also smiled and said to himself, "That is the Hassan I know."

Chapter 14: New Beginnings

"Cry. Forgive. Learn. Move on. Let your tears water the seeds of your future happiness" - Steve Maraboli

Everyone left the room, leaving behind all of the events of the day. Hassan hadn't smiled this much in months. John, Kyle and Cara were driving. Seeing as Andrew had a decent location to go, John followed him. Andrew was accompanied by Pete, Hassan, and Ethan. Andrew was driving while Hassan sat next to him. He removed his shoes and he sat down with his feet on the seat, opening half of the window. When Andrew started playing some songs, Hassan began singing along with them while staring out the window. Andrew could feel Hassan breathing for the first time after he pointed out what had transpired between him and John.

After this, he couldn't tell John what Brian had done because it would ruin him and Cara. Both Andrew and John parked and joined the others in walking to a local pub located downtown. They were seated at a table and ordered drinks. Having approached the bartender, Hassan and Andrew requested several drinks. When Hassan finished taking Andrew's order, Andrew headed to the bathroom. While Hassan was waiting, a man approached him and began to make advances. Hassan revealed to the man that he is HIV positive. Hassan was left alone as the man became alarmed and ran away. After seeing his reaction, Hassan burst into laughter. "What a lousy pick-up line!" remarked a guy to Hassan.

Hassan, at first, stood in awe as to what a stranger had just said to him, then replied, "Excuse me?" The stranger continued, "Hi, I am Klaus, and you?" Hassan was intrigued by the stranger's unique demeanor, so he tried to carry on the conversation, "Hassan, nice to meet you, Klaus."

"Likewise. Allow me to say that you have a terrible pickup line."

"He hit on me, so I told him the truth."

"You have a beautiful smile."

Hassan struggled in retorting around Klaus as he was too good for Hassan to handle.

"Is that one of your pickup lines?"

"Yes, but in your case, it is true."

"Thank you, I have to go back to my friends with the drinks."

Hassan attempted to escape from the conversation but felt confined again.

"Is one of them your boyfriend?" Klaus asked.

"You don't give up, do you?"

"No." Klaus was determined.

"I am single, and they are my family," Hassan responded.

Hassan walked toward the table and put the drinks on it. Klaus followed Hassan as he wanted to talk to him some more.

"So, you are here with your family, not to pick up guys."

"I don't go to the bar to pick up guys. This might come as a surprise to you, but I've been with only two guys in my entire life. Both of them were my boyfriends."

"You are telling me that you never had a one nightstand?"

"Never."

"How is that possible?"

"Why is it so hard to believe?" Hassan chuckled.

"You are good-looking and have gorgeous eyes. Are people blind?"

"Is that another one of your pickup lines?"

"Believe me, it is not."

"So, you are going to bars a lot?"

"Almost every week for the last three years."

As Hassan came to know about Klaus' intentions, he wanted to inquire more about his life.

"Wow, that is a long time. So, you are looking to hook up, nothing serious?"

"I was in a relationship with a man for three years, but he left me in pieces."

"I am so sorry; I didn't mean to bring it up."

"Would you like to spend the night with me?"

Hassan was taken aback by what Klaus had just asked him and did not want Klaus to have any hopes.

"Look, Klaus. I am sure you are a nice guy, but I am not that kind."

"Why not, try me." Klaus insisted.

Klaus moved close to Hassan, held him between his arms, and kissed him.

"What do you think now?"

"That was nice, but I am not going to say yes." Hassan kept his stance.

"Come on, I am not a bad guy." Klaus tried to convince him.

"You are not. Look, I am hungry. Do you want to grab a bite with me?" Hassan was trying to avoid the situation, which was getting rather tense.

Klaus was more than happy to avail this opportunity, while Hassan was getting annoyed.

"I cannot say no to this smile."

"Stop it. I am not going to fall for your lines."

"Shame."

Hassan laughed and replied, "Get over it already."

When Hassan returned, he went to tell the others that he would be going out for a little time. Afterward, he stepped outside with Klaus to grab a bite to eat. They proceeded to a neighboring shawarma stand and got some wraps. All this while, they continued to eat, talk and joke around with each other. Suddenly, Klaus paused for a moment and grinned at Hassan.

"Are you okay?" Hassan inquired.

"I haven't enjoyed a man's company in a long time. They all jumped to sex directly," Klaus confessed.

"You are a person, not a sex machine." Hassan said nonchalantly.

"Can I ask you how did you get HIV?"

"It's a long story for another time." Hassan stalled Klaus' query.

At that moment, Klaus received a call; he asked Hassan if he could answer it and walked away from him. Suddenly, he started to yell and seemed worried. Hassan walked to him, took the phone from his hands, and talked to the person on call. Hassan ended the call and said, "You have a child!" in utter shock.

"Tell me, is he okay?" Klaus expressed his concern.

"He is okay."

Klaus stood up and wanted to leave.

"I have to go."

"I am coming with you. You are very emotional right now. You might say things you will regret later on."

"Okay. Come with me, I have to take my car first."

Klaus took Hassan to his car and drove home. "What is it about Hassan that makes me feel so comfortable? "Who is this man?" Klaus was lost in his thoughts. Suddenly, Hassan interrupted by asking, "Is the situation under control?" Hassan asked Klaus. "I hope so," Klaus answered. They finally made it to Klaus's place. They both entered the house together. Klaus became enraged as soon as he entered the room and began yelling. Because Klaus was furious, the nanny was unable to speak with him. Hassan told him to calm down and that he would talk to the nanny about the situation later on. Hassan was informed by his nanny and found out that the child was in good health.

"Did you hear? Your kid is okay, now calm down." Hassan tried to comfort Klaus.

"My kids are all I have." Klaus sobbed.

"You have more than one! How many do you have?"

Klaus pointed with his hand as four.

"Wow, I never saw that coming. I want one, and you have four. That's not fair."

"Now I remember you; you are that guy from the shopping center." The nanny pointed to Hassan.

"You are that lovely lady. I believe your name is María." Hassan also remembered her.

"Yes, my drear. I've never forgotten your face."

"How is your son?"

"I did what you said to me, and he came back."

"I am happy for you."

"Do you know each other?" Klaus intervened.

"We've met once."

Hassan kissed María on the forehead. Interested, Klaus asked himself, "Who is this guy?"

Hassan, who took the infant from Klaus, chatted to him and swayed until the baby fell asleep before putting him in his crib. Klaus stood right behind Hassan and kissed him as he turned around. Klaus kissed Hassan again, and Hassan kissed him back. Inhaling deeply, he placed his head on Klaus. Then, Klaus asked Hassan again to spend the night with him to which, Hassan nodded in agreement.

"Can you spend the night with me?"

"Yes."

"What has changed?"

"When you kissed me, I didn't feel that pain in my chest and rash in my veins that I usually feel in a first kiss. It's like my body knows you, wants you. I can't explain it."

"You need to know I never have…"

"Really!" Hassan interrupted Klaus with excitement.

"Yes, that is right. One more thing, I won't have protected sex with you."

"No, I won't do that."

"Are you undetectable?"

"Yes."

"You are undetectable, and I am on PrEP. I don't see a problem."

"No." Hassan broke into tears.

"Have no fear. There's no need to worry about anything." Klaus took Hassan to his room. They immediately started to kiss each other. They were unable to stop. They both undressed each other. As Klaus pulled Hassan onto the bed, he wrapped his legs around Hassan's body and kissed Hassan passionately on the lips for a long time. Klaus bit the leg that was pressed on Hassan's chest as he shifted it. Hassan's leg was licked from head to toe by Klaus. Seeing Hassan, Klaus reached inside of him. Klaus was driven to distraction by Hassan's displays of pleasure and anguish. He kissed Hassan again, then he stopped, put his hand around Hassan, and leaned in.

"We should do this often."

"I thought it was supposed to be one night."

"You said it yourself. Your body knows me, wants me."

"I need my phone. I have to text my friends." Hassan tripped suddenly.

"Okay."

"I'll be spending the night with a buddy," Hassan informed Andrew via text message. Klaus kissed Hassan on the cheek as he sat behind him in bed. Hassan moved to the same position as Klaus, who was now seated on his knees behind Hassan. Having forced his way inside Hassan, Klaus put his arms over his chest. Klaus moved into Hassan as he kissed him. It seemed as though Hassan's physique was known to him. As Klaus finished, he pushed Hassan onto the bed, where they continued to make out. After a while, Hassan and Klaus stopped and stared at each other.

"Who are you? Why do I feel like I know you?" Klaus was confused about his feeling of belongingness to Hassan.

Klaus was kissed by Hassan as he grinned and rubbed his cheeks. The two fell asleep while Klaus held Hassan in his arms.

The next morning, Hassan and Klaus were talking when a dog entered the room and sat between them. Hassan grinned at him as he found his way to Hassan's lap.

"Hey, buddy. What is your name?"

"Her name is Bella."

"Nice to meet you, Bella."

"We need to wear something. They are about to attack us."

"Attack us? What is going on?"

"Just put on your clothes, and I will explain later."

Hassan and Klaus got dressed quickly.

"My kids, every time I bring someone to stay overnight, they attack him in the morning to make him leave."

"I see. Do you want me to leave?"

"No, this time, I want you to stay."

"Wait, I heard something."

"Are they planning something?" Hassan asked. Both of them waited behind the door, listening carefully. In response, Hassan grinned and asked Klaus to lie down on the bed. Suddenly, a youngster, approximately eight years old, walked into the room and glanced at the bed; no one else was there. A man in the background called out, "Gotcha." Holding the kid in his arms, Hassan placed him on the bed. Klaus introduced his kid to Hassan.

"This is Adrien. Adrien, this is Hassan."

"Nice to meet you, Adrien."

"You have lovely eyes, and I like you," Adrien complimented Hassan.

"I think we can be friends." Hassan said to Adrien. "As your friend, I warn you. They will scare you." Adrien said to Hassan.

"Can you keep a secret?" Hassan turned toward Adrien.

"What is it?" asked a curious Adrien.

"I don't get scared easily." Hassan reassured the kid.

Adrien laughed and said, "You are funny. I like that."

"Who wants pancakes?" Klaus suggested.

"Don't eat the pancake. It is not good; he doesn't know how to cook." Adrien warned Hassan of Klaus' cooking while whispering in his ears.

"I can cook for you," Hassan whispered back.

"That would be nice."

Hassan was given a kiss on the cheek by Adrien. Hassan's popularity shocked Klaus. They all walked to the kitchen as he grinned. It was Hassan who was in charge of cooking pancakes. He paused when he felt something.

"Is everything okay?" Klaus asked, concerned.

"What is that smell?" Hassan asked him in return.

"It is the pancakes," Klaus replied, confused at his question.

"Do you have anyone pregnant at your house?" Hassan tried a banter.

"Did you tell him?" Charley asked Klaus.

"No one told me a thing. You are glowing like a rare flower. You are definitely carrying a girl." Hassan said to Charley as they introduced each other.

"How do you know?" asked a curious Charley.

"I am Hassan."

"I am Charley. That pancake smells good, but I cannot eat it. The baby wants something different."

"Are you all up for breakfast in a restaurant?" Hassan asked everyone present around him.

"Yes, which restaurant?" Charley asked before Klaus could

"Wait, and you will see. Klaus, can we all go together to my restaurant?" Hassan replied to her then turned toward Klaus.

"You own a restaurant?" Klaus asked, surprised.

"Yes, I do."

"I will change and will inform Edouard."

Charley went to change, so did Adrien.

"Charley, Adrien, Edouard, and the baby is?"

"Adam." Klaus replied to Hassan.

"What a lovely family. I think your car is not enough. I will tell you the address. I will take María and Adam and Adrien with me."

"Text me the address," said Klaus.

"If you want my number, just ask." Hassan replied while smiling.

"Can I have your number?"

Hassan handed Klaus his cell phone after opening it. Hassan's phone had Klaus's number saved on it so he sent him the address through text message. "This is for good luck," Klaus told Hassan as he kissed him. Then, Hassan gave Klaus a long, passionate kiss and whispered, " This is to think of me." Klaus grinned as he placed his palm on his lips, glanced at Hassan, and looked away. Everything seemed to be moving slowly to Klaus. He stopped Hassan in his tracks and kissed him on the neck as he was ready to walk away. Hassan grinned and began to walkdown the hall.

Hassan got an Uber to pick up María, Adam, and Adrien while Klaus drove Charley and Edouard. They arrived at their destination at the same time. Klaus was parking the car when Charley and Edouard got out. Hassan held Adam in his arms.

"So, you are Edouard. I am Hassan." Hassan introduced himself first.

"Nice to meet you, Hassan," replied Edouard.

"Nice to meet you too, Edouard."

"The smell is fantastic. What are we having for breakfast?" Charley was eager about the breakfast menu.

"You can see the menu and decide," Hassan told her.

"I can't wait," she said excitedly.

"Everyone go inside; they are expecting you."

As María and Adrien walked into the restaurant, Charley and Edouard joined them. Klaus was waiting for Hassan, who was holding Adam in his arms. Adam and Hassan were playing. Klaus grinned at them. He thought back on how his ex-husband treated his children and felt grateful to have met such a unique individual. He looked at Hassan cradling his infant and he was in awe of the scene in front of him. Hassan's smile reminded him of something, but he couldn't recall what it was about. "Let's have fun!" Klaus said to Hassan as he kissed him.

Hassan and Klaus entered the restaurant together and were welcomed by the rest of their family. Hassan was still holding Adam. He even turned down María's offer to take Adam, saying, "You sit down and enjoy your breakfast. I will take care of Adam." Everyone was watching Hassan and Adam's game. Adam was in Hassan's sights, and he hugged him near to his heart and smelt him. His lengthy sigh was followed by the words, "I want a baby." To this, Charley said, "You can take mine." Everyone was surprised by Charley's comments.

"What?" Klaus intervened.

"Klaus, you need to stay out of this. She was talking to me." Hassan tried to calm him down.

"She is my daughter, and you cannot take my grandchild from me." Klaus looked angry at the suggestion Charley had made.

"No one will take anyone, but I need to talk to her in private," Hassan said politely.

Hassan turned to Charley and asked if she would want to join him. He took Charlie to one of the restaurant's workers after he handed Adam to María.

Hassan introduced the restaurant lady to Charley. "Her name is Ruby. She is 24 years old. She is pregnant with a boy. She is a single mom, but she never gives up. Being a single mom comes with a unique set of emotional challenges that could, at times, feel overwhelming. It's like walking with a heavy backpack that weighs you down sometimes, and when you take it off, your body starts to crave that weight, and you feel anxious until you can put it back on. Do you think that we need Ruby to work here with a big belly? We make her supervise nothing more to support her. She deserves everyone's support. She is a powerful lady, so are you. You are not alone. You will have all the support you need. You don't need a man to define you. You are a strong, powerful woman."

"Will you be there?" Charley couldn't stop herself from crying.

"All the way long," Hassan assured her.

"Promise?"

"Promise." Hassan hugged Charley after giving her assurance of his support.

"I want ice cream," Charley said.

"After you finish your breakfast."

"Deal."

They strolled back to the table, where the rest of the family were seated. Hassan glanced at Klaus in confusion. He felt he wanted to speak to him, so he asked him to join him and started walking.

"What did she say?"

"She will keep her baby."

"What did you say to her."

"I showed her what she needs to see. You need to get more involved in your kids' life."

"I don't know how to get more involved. Since their mother died, I am hopeless. I don't know what to do."

"You did a fantastic job of raising three wonderful kids."

Klaus was in tears when he replied, "She is twenty-three years old and pregnant. Edouard is a genius but not doing well at school. Adrien is the only one who tells me everything. I lost their mother nine years ago, and it's been so damn hard since."

"Don't beat yourself. You did your best."

"My best sucks," said disappointed Klaus.

"It is hard to be a single father for three kids. But you did your best without anyone's help." Hassan tried to comfort him.

"You are so young for that," Klaus said, wiping his face.

"Excuse me! How old do you think I am?" Hassan asked him.

"Twenty-seven years old!"

"I am about to turn thirty-six," Hassan laughed as Klaus guessed his age wrong.

"What? You look way younger." Klaus was surprised while Hassan asked Klaus his age with a smile, "How old are you?"

"I am forty-seven years old. Is that too much for you?"

"You are eleven years older than me," Hassan said.

"Is that bad?"

"Age is just a number, and basically, you are a teenager."

"What do you mean?"

"The man I met yesterday was full of teenager's energy."

"Do you want to see what this teenage man is capable of?" Klaus was in a mood to display his energetic personality to Hassan.

Hassan laughed and Klaus gave him a long passionate kiss. Hassan looked up at Klaus's face and smiled.

"You are the only person who took me on a date with my kids. I am glad I met you." Hassan was overwhelmed with his date at Klaus'.

Hassan replied, "Me too".

Chapter 15: Breakthrough

"True love asks no questions, makes no reservations, but puts itself unconditionally in to the hands of the loved one" - Paula Marshall

Soon, Hassan and Klaus returned to the table. Edouard kept his distance from Hassan but, at the same time, he had a close eye on him. He appeared to be upset. Hassan noticed his behavior and asked Edouard, "You didn't like the food?" He looked over at Hassan and told him that everything was fine. Laughing, Charley punched Edouard under the table. "I'm sorry," she said, "he's a little cranky in the morning and on weekends." Before Hassan had a chance to respond, his phone started ringing, so he excused himself.

"What are you doing?" Charley asked Edouard.

"I am sticking to our plan," Edouard replied calmly.

"We finally found a good guy who treats us well, and you decided to stick to our plan? Do you want dad to find another Jeffrey?" Charley confronted him.

"I don't want anyone to take my mother's place," Edouard said, now seeming a little sad.

"No one ever will take mom's place. Did you notice how he treated us? Or the way dad smiled? Have you ever seen him smiling like that? Let him be happy," Charley assured Edouard.

"Shhh. He is coming." Adrien tried to keep everyone quiet.

"Why is everyone silent?" asked an intrigued Klaus, who tried to probe into the matter.

"Will you see Hassan again?" Charley diverted Klaus' attention.

"I don't know, but I would like to."

"Good, I want to be able to come here from time to time. I will also tell my friends about this place." Charley beamed with excitement.

"All you are thinking about is food. Give it a rest." Edouard made a rather futile attempt to stop Charley from excessive eating.

"Get pregnant, and you will know." Charley retorted. Edouard was opened his mouth to answer her but right that moment, Hassan joined them back at the table.

"Here is what you asked for," Hassan handed Charley the ice cream.

"You are an angel. Thank you."

"My pleasure. Anyone has a special request?"

"Can I get ice cream?" Adrien requested.

"Yes, you can. What about you, Edouard?" Klaus asked.

"Ice cream is fine." Edouard said.

"What about you, María?" Hassan asked.

"I just want a coffee if it is possible."

"Ice cream and coffee coming right away." Hassan got back up to notify the kitchen staff.

Klaus walked with Hassan and said, "You asked everyone, but you didn't ask me." Hassan stopped, looked at Klaus biting his lip, and replied, "I have a different plan for you." He whispered in Klaus' ears. And they both laughed. Klaus watched him walk away with a smile on his face.

Once everyone was done, Klaus took the kids and returned home. On the way home, he asked them, "What do you think about Hassan?"

"I love him. He is super nice," said Adrien.

"He is different from anyone you have dated," Charley approved of him.

"What about you, Edouard?" Klaus asked him as he hadn't spoken a word since they got in the car.

"Did you pick him up from a bar or a street?" Edouard asked.

Klaus clenched his teeth and hit the steering wheel with his hand. By the time they reached home, Klaus' mood had changed; he walked straight into his office and locked himself in.

Charley, who could see the disappointment in her father's demeanor, turned to Edouard, giving him a stern look.

"What?"

"Nice job," Charley said sarcastically.

"I did nothing wrong," Edouard couldn't see what he had done wrong.

"When you will understand that mom is gone. He is lonely. We are what he is left with. He finally found someone different from everyone he has dated and makes him happy. Stop ruining everything. For our sake, please stop. Don't you see we are already a broken family?" tears started to form in her eyes.

Hearing this, Edouard ran to his room to hide. He picked up his mother's picture from the nightstand and held it in his arms. He fell on the bed and began to cry.

Adrien looked at María and asked her, "Why can't we just be a happy family?" María hugged Adrien and told him, "It is a matter of time."

As Hassan returned home and opened the door, everyone turned their attention to him. "Good afternoon," Hassan said as he removed his shoes and entered the living room. Everyone bamboozled Hassan with their questions about Klaus.

"Did you have sex last night?", Andrew asked instantly.

"Who is he?" Ethan jumped in too.

"Tell us every detail," Cara was interested in the story too.

"I thought you were not into one night's stands," John said in doubt.

"Come on, say something," Kyle insisted.

"Yes, please," Pete seconded Kyle.

Hassan walked to the couch, sat down, and turned on the television. 'You won't say a word,' Andrew told Hassan.

"Do you remember the time when we went together to the shopping center and a nice lady spilled hot coffee on my shirt?" Hassan laughed and started talking.

"Yes, but how is that related?" Andrew seemed confused.

"The guy I went with last night was Nicholaus."

"Who is Nicholaus?"

"No way," Andrew was in denial.

"Yup, I slept with Nicholaus," Hassan tried to clear the doubts.

"Who is Nicholaus?"

"A guy that the nice lady works for. Remember who spilled the hot coffee on Hassan's shirt months ago? That lady wished that Hassan and Nicholaus should see each other at that time, and now her wish came true," Andrew narrated the story quickly.

"Whatever, talking encyclopedia," Kyle did not acquiesce at all.

"Sorry, who is Nicholaus?" Ethan still could not fathom what Andrew had said.

"The guy who Hassan slept with," John verified the story.

"Is he French?" Cara further inquired.

"Yes, as a matter of fact, he is," Hassan nodded.

"Isn't he old?" John asked, sounding a little jealous.

"I dated you, and you are older than me," Hassan told John.

"Dating someone older than me is the best thing that has happened to me," Pete said.

"Age is just a number, and eleven years are not that a big deal. Anyway, it was nice." Hassan said.

"Good for you," said Andrew.

"I need to take a shower because I feel dirty," Hassan excused himself from everyone.

"Oh, daddy, can't get enough of that honey."

Andrew's head was hit with a pillow thrown by Cara. Afterward, Hassan went to his room to shower and change. Hassan checked his phone after getting out of the shower. He had received a message from Klaus that said, "Can I speak with you?" Hassan called right away and he answered.

"Hey, are you okay?" Hassan asked, worried.

"Can I come to your house?" Klaus asked without answering his question.

"Are you crying?" Hassan asked again.

"I was."

"What happened?"

"Can I come to your place and talk?" Klaus insisted.

"I will send you the address."

Hassan quickly dressed and went downstairs. Andrew saw coming down the stairs in a rush and asked him,

"Are you alright?"

"Klaus is on his way. He wants to talk." Hassan told him.

"Who is Klaus? Oh, Klaus. Daddy can't get enough?" Andrew said with a suggestive smile.

"He was crying," Hassan said.

"You are that good?" Andrew continued teasing Hassan.

"I am serious."

"If he wants to talk, then talk to him. I don't see a problem." Andrew said, now in a serious yet comforting tone.

"I am worried if John saw him." Hassan expressed his concern.

"I see your point, but he has to get used to this kind of situation," Andrew assured him.

"Okay."

Klaus arrived at the house and entered. Everyone was keeping an eye on him.

"He is handsome," Pete complimented Klaus.

"Eh, I am here," Ethan interrupted Pete.

"I think that means I have to keep my mouth shut. You are the most handsome man ever," Pete whispered.

"Does Hassan date only tall people?" This was Andrew.

"He looks nice," Cara said.

"I agree," John agreed with Cara.

"Nice job, Hassan. He is yummy," Andrew teased Hassan again.

Hassan took Klaus to his room to talk. "Sorry about my friends staring at you. You wanted to talk?" Hassan said.

"You said I have to get involved in my kids' life." Klaus started to speak.

"Yes, I did."

"I don't know how to get involved in their life. It seems like they have built a wall that I can't breakthrough. I feel the space between us growing deeper and much darker every day. I am hopeless and can't do anything," Klaus burst out in tears.

"No one knows how to do it. It is not easy. But you can start by becoming their friend and then take it from there," Hassan tried to convince Klaus.

"Can you please help me?"

"Why do you think I can help you?" Hassan asked him.

"What you did today with Charley, I would never be able to do it. You are a family person, and I can't feel I have a family."

"Don't say that. If you need my help, I need to know how you drifted apart from your family? What happened?"

"I think it all happened when I was in a relationship with a man named Jeffrey. He didn't like my kids, and I was blindly in love with him. I hurt them deeply in a way they couldn't forgive me. He destroyed my family and shattered me into tiny pieces. I tried to pick up those pieces as much as I could and put them back together. These pieces of me were forgotten by time, and I lost myself. I don't want to lose them again. I am not a bad person," Klaus could not stop crying.

Hassan hugged Klaus and said, "You are not a bad person. You are a father who is trying to do everything he could to get his family back. To be honest, I can't help you, but I can tell you what you have to do."

"I am listening," Klaus said, wiping his tears.

Hassan told Klaus so many things that he could do to earn back his children's respect and walked him through the process of getting his family together again. Klaus was paying close attention to every word Hassan was saying. He looked at Hassan and asked him a question when he stopped. "How do you know these things?"

"I am too many things, don't underestimate me," Hassan said obnoxiously in an effort to cheer Klaus up.

"You asked me things; it only can work for every one of my kids separately. How could you ever think like this if you are not a parent?"

"I learned everything from my parents."

It was clear to Klaus that Hassan was in pain when he spoke of his parents and that his eyes were wet. In his arms, he cradled him.

"Can I ask you where you are from? I still cannot recognize the accent you have."

"I was born in Iraq. My father is an Iraqi, but my mother is Turkish."

"It is fun to look at your face in the morning and see those eyes. It's the best way to wake up." Klaus changed the topic to ease up Hassan's mood.

Hassan whispered in Klaus's ear, asking him to continue talking about anything and everything. He slowly walked to the door and suddenly opened it to find Andrew standing on his tiptoes. Others looked embarrassed when Andrew fell to the ground. Hassan asked, "Can I be of assistance to you?" And they all left immediately, without saying a word. It was nice to see

Hassan and Klaus laughing together, and Klaus said, "You are mean." Hassan said with a smile, "I know."

"They all love you."

"They are the only family I have," Hassan said with a smile.

"The one with sad eyes is your ex-boyfriend, right?" Klaus asked him.

"Yes, how did you know?" Hassan was surprised at his guess.

"The way he looked to you and me. I want to ask you something."

"Sure."

"Since you told me you have been with two guys, I suppose you got it from your boyfriend. My question is, how on earth did a person like you got cheated by his boyfriend?"

"Long story short, he is of a very jealous kind. One day I got sick, and someone took advantage of me while I was under medication. He took photos and sent them to him. He couldn't tell me, so he started sleeping around." Hassan could not hold on to his tears while sharing his side of the story.

"I wish I'd met you instead of meeting Jeffrey. At least my life would be different now."

"That's not even possible," Hassan said.

"How?"

"You met Jeffrey even before I came to Canada."

"I said I wish," Klaus said, rolling his eyes at him.

"Okay."

"What did your friends say about me?" Klaus asked.

"Well, they call you daddy," Hassan replied quickly.

"Am I that old?" Klaus said with concern in his voice.

"I can't even call you that."

"Someone promised me a treat." Klaus remembered their conversation from the restaurant.

"I did not." Hassan quickly retaliated.

Klaus dragged Hassan to lay down on the bed next to him.

"I have a very long day tomorrow," Hassan said.

"Why is that?"

"I declared in front of the whole company that I am living with HIV."

"I hope that won't be an issue," Klaus assured him.

"We will see about that tomorrow."

Klaus embraced Hassan and closed his eyes for a few moments before returning to his house.

After seeing Klaus off, Hassan went around the house looking for his friends. He heard voices coming from the basement, so he headed there. He opened the door and saw them huddled together. John was the only one who wasn't laughing. Andrew looked up at John and placed a finger on his hand. John was surprised.

"He's moved on," said John.

"Stop looking at the past, hold yourself, move on, and don't ever look back," Andrew told him.

"I don't know how."

"You will find the right person, but don't make the same mistake again," Andrew assured him.

A tear fell out of his eye. "It's okay," Andrew said as he put John's head on his chest, "you should let it go now." John and Cara sat side-by-side, and she held his hand. On the floor, Ethan leaned against John's legs and placed his hands on them. Archie sat on John's lap, while Kyle sat next to Cara and Pete sat next to Kyle. It was then John blinked his eyes open, embraced Archie, and kissed him. John smiled when Archie licked his face.

The next day, John and Andrew left for work together in the morning. John bought two cups of coffee at the coffee shop on the way. John ignored the barista's smile, like he always did, and continued to ignore him. In the meantime, John sipped his coffee and chatted with Andrew in the yard. As John was sipping on his coffee, Andrew asked him, "Who is Carlos?" while looking at the cup. "Why?" asked Andrew. Instead of replying, Andrew showed him his cup.

"It seems someone has a crush on you," Andrew said.

"Do you know him?" John asked.

"I think he is the barista guy. He wrote his number on the cup, why don't you call him sometime," Andrew suggested.

"I can't do that."

"Why not? I will have a look at him through the glass. Just tell me which one he is." Andrew said to John.

Seeing Carlos, John and Andrew drew near the window.

"Wow, you should hit that."

"He is like twenty-one years old." John settled back in his seat.

"Daddy has an issue with that? He is an adult." Andrew teased him.

"I am living with HIV." He reminded Andrew.

"You are like a broken record. Maybe he is open-minded. It's worth a try. Give me your phone." Andrew said excitedly.

"What are you doing?"

"Open your phone and give it to me." He insisted.

John handed Andrew his phone after he opened it and dialed Carlos' number for him. Carlos politely responded.

"Hey, Carlos. I am John. You wrote your number on my coffee cup. I am wondering if you would like to grab a drink with me." Andrew said to him without giving John a chance to understand what was going on.

"Hey, John. I would love to." Carlos said cheerfully.

"Great, you have my number. Give me a call when you are ready. I have to go now. See you."

"See you."

Andrew ended the call and turned to John to discuss Carlos' situation. John's phone call had made Carlos happy. John thanked Andrew for saving Carlos's number, who returned the phone to him. When Andrew saw John, he said, "Don't screw it up or I'll kick you in the back."

Meanwhile, Tristan was asking around in the office, "Have you seen Hassan since he arrived at work? He had to finish some documents before noon." Hassan was working at his desk when Tristan passed by. He walked up to the corkboard and smiled as he did so. Took a picture of himself with Hassan and went to his desk to talk to his coworkers.

Hassan was photographed by all of them. They pinned the photo to the corkboard and left it there. Klaus called Hassan to see how he was doing at work. Hassan decided to take a walk as he talked on the phone. His attention was drawn to the corkboard, which he passed before stopping. He told Klaus he would call him back shortly. Seeing the contents of the board, tears started streaming down Hassan's face. He walked up to the corkboard and stood there for a while. He was unable to speak a word. He snapped a picture of the corkboard with his phone and put it in his pocket. He then walked away. Putting his arm around Hassan, Tristan said, "Everyone loves you." "You're a good person."

"Thanks," said Hassan, he said as he looked at the corkboard. Then Hassan went back to his desk. Klaus received the picture from him. Klaus smirked as he saw the picture in front of him. His text message to Hassan read: "This is a good day for you."

***

Carlos sent John a text message asking, "What about 5:00 pm?" John texted back, "Sounds good. You can meet me outside the coffee shop." To this, he replied: "I'll be on time." Although John was apprehensive about meeting Carlos, the meeting was necessary for him to move forward. It was 5:00 p.m. The two met up and decided to go for a walk.

"Finally, you and I on a date," Carlos said happily.

"What do you mean by finally?" John was confused.

"I have been hitting on you for months," Carlos opened up.

"Oh." John finally realized it.

"So, here we are."

"Carlos, I am sure you are a nice guy. But I am not sure if I am good for you." John decided to talk to him openly.

"If you are talking about age, I don't mind," Carlos assured him.

"I am living with HIV," John told him.

"I see. Are you taking your treatment?" Carlos didn't seem fazed by it.

"Every single day."

"If you are HIV+ and taking treatment and maintain an undetectable viral load, you can have sex knowing that you won't pass HIV to your sex partner. In short, when HIV is undetectable, it's untransmittable. I am studying biochemistry. It's not a reason to prevent yourself from having a normal life," Carlos ensured John.

"You are telling me that you are okay with it. Oh, my God. I spent the last few hours freaking out." John said, sounding relieved.

"No need. Now can I get my kiss?" Carlos asked him.

John smiled and kissed Carlos on the cheek. Carlos smiled at John before saying, "Where are we going?"

"Are you hungry?" John asked.

"Starving."

"Pick a restaurant."

Carlos said, "I'll take you to one of my favorite restaurants." As they walked, John smiled and joined him.

As dinnertime approached, everyone at home gathered around the dining table to enjoy the meal. Except for John, everyone was present.

"Where is John?" Hassan asked.

"He is on a date." It was Andrew who answered.

"Really!" Cara was surprised.

"Anyone we know?" Ethan joined in.

"With a hot Latino. Daddy needs to hit that town; you know what I mean." Andrew disclosed everything.

"We all know what you mean, but I don't think John will do it on the first date," Ethan said.

"Oh! So now that will be Hassan's thing? Hmm, interesting!" Andrew said to tease Hassan.

"I am still here." Hassan glared at him.

"Hassan, you owe us details." Kyle insisted.

"We are dying to know." Pete partnered up with Kyle.

"It seems he is into you." Ethan also shared his opinion.

"We met, and it was probably a one-night thing. It was just nice." Hassan said.

"A one-night thing! You spent the whole night and the next morning with him, and then he came here, and you are telling us it was a one-time thing?" Andrew asked.

Hassan put his phone next to him on the table. At that moment, he received a text from Klaus.

"So, you won't see him again?" Ethan asked.

"I will see him next Friday night. He just texted me." Hassan said with a smile.

"What? Are you trying to fool us? You are into him." Andrew caught Hassan's lie.

"I don't know why, but I feel I know him, and my body knows him. I didn't want to resist him." Hassan finally shared his feelings.

"Poor child, one night with him, and you are falling for him." Andrew just won't stop teasing Hassan.

"What do you think? I have been with two guys in my entire life, and someone made my whole world crumbling under my feet. Now I just got back on my feet, and I feel lost again. I just don't know what to do." Hassan started crying.

"We know you have dealt with a lot of things, but we are all here for you." Cara showed her support to Hassan.

"Do you know how many people love you, how many are willing to support you? Even if you are lost, you will always find your way. Did you see what they did today at the company?" Ethan asked Hassan.

"I did."

"What did they do?" Andrew asked.

"Ask Hassan," Ethan told Andrew.

Andrew received a picture of the corkboard from Hassan on his phone. Cara took the phone from Andrew to take a look. Almost everyone looked at the image. Hassan was greeted with a smile and a look of admiration.

"Now we need details," Cara said.

"Do you want the long version?" Hassan asked.

"Everything, but skip the sex part."

"Okay."

Then Hassan began to describe Klaus and what transpired that night and the next day. Everyone was eating and listening at the same time, and it was quite entertaining. The group laughed as usual when Andrew made a series of funny remarks.

After dinner, John had returned from his date. He was excited and happy. Everyone sat in front of the television and watched their favorite show. John stood in front of them, beaming with pride. Hassan turned off the TV. He was tired of watching it.

"Someone is happy." Andrew broke the ice.

"My date was great," John said excitedly.

"Yes, we can see that. And we are happy for you." Hassan said.

"Did you talk to him about your fear?" Andrew asked.

"I did. He was very understanding. He is more like…"

"More like what?" Cara interrupted.

"More like Hassan." John sighed while replying.

"Is that kind of insult? Or what?" Pete asked as he was confused.

"He is very spontaneous, full of life, and tasty," John replied, laughing at his question.

"You mean, you found another Hassan?" Andrew asked.

"What he is trying to say is that he is his type," Hassan explained to Andrew.

"Thank you," John said to Andrew.

"You are welcome. Now you owe me a boyfriend."

Andrew replied to John.

"You are mean," Hassan said to Andrew.

"I know."

Everyone in the room roared in laughter. Everything that had happened with Carlos was recounted by John. Hassan was happy for him. When Hassan's phone started ringing, the conversation was interrupted. He picked up the phone and went upstairs to answer it. "Nicholaus is calling. Hurry up," Andrew said to everyone.

"Hey there," Hassan said, picking up the call.

"Hi, how are you doing?" Klaus asked.

"I am doing great, thanks. What about you?"

"I am doing good. I did what you told me to do. Now I am about to move to the next step."

"Good luck with it," Hassan said assuringly.

"Thank you. Did you get my message?"

"Yes, I did."

"I am about to take a shower. I wish you were here. Daddy is hungry right now." Klaus teased him.

"Obviously."

"I have to go. I will text you the details." Klaus said goodbye to Hassan and went for a shower.

"Good night."

"Good night."

Klaus smiled as he ended the call and put the phone on his chest. He found Edouard was asleep in his bed when he entered Edouard's room. As it turned out, Edouard wasn't sleeping at all! He was pretending. Klaus sat beside him and began to speak. "When it came to standing up for you, I was blinded by emotion. I'd like to go back in time, but it's not possible. I would do everything differently. I would choose you over anyone else and would never allow anyone to cause you harm. You're my child, and I adore you without reservation." Klaus kissed Edouard and bid him good night. Edouard wept after Klaus left the room and said, "I love you too, dad." As Edouard closed his eyes, he drifted off into a deep sleep. It took him some time to open his eyes, but eventually, he did and walked straight into his father's room but Klaus wasn't there.

As a result, he searched every room of the house for him but was unable to locate him. He entered a room with a water-covered floor. He took a single step and fell into a large pit. On the ground was a layer of water. During his wandering, he came across a book lying on a wooden chair.

His eyes widened as he approached the chair, picked up the book, and began to read. In addition to pictures of him and his family, the book contained words written in a foreign language that he did not recognize. He studied the pictures and went through each page, examining it for a long time. It dawned on him that the pictures were a record of his memories. Jeffrey brought back many memories of his father for him. He became angry and tore the pages out of the book before throwing it on the ground and crying out in pain. He then found unlit matches in a matchbox lying on the chair.

He picked it up, lit a matchstick, and threw it on the ground before putting it back down. Starting on the floor, the fire then moved up and onto the book. Along with everything else in that hole, the book began to burn. He turned around to find his father cuffed to a wooden pole with his hands and feet tied. His father was on fire when the flames reached the wooden pole. But no one could hear him as he screamed for help. 'I wish you were here,' he sobbed. As if on cue, the flames went out. Nothing was left when he looked around. He could see a shadow walking toward him; he knew it was his mother.

He was unable to move or speak a single word. "Don't let your hatred blind you, my child. All that you have will be taken away from you because of hatred. As a result of hatred, your family will burn to the ground. You must give your father a chance to show you that he cares for you. Don't get in his way. I'll always be there for you," his mother said. As Edouard closed his eyes, he wished to be awoken from his dream. Opening his eyes, he discovered that he was lying in his bed. It was then that he looked at his mother's picture, grabbed it, and held it near his heart as he fell asleep again.

The next morning, Edouard washed up and changed into his school uniform before heading off to school. He smiled as he went to the kitchen to prepare his breakfast.

Chapter 16: Unanticipated Rifts

"The weak can never forgive. Forgiveness is the attribute of the strong" - Mahatma Gandhi

As Edouard was preparing his breakfast and getting ready, everybody in the house stared at him in astonishment, "Where is that sad face of him?" was the question everybody had in mind. He ate his breakfast and got ready for school, but he was late. Inquiring about his friend's availability after school, Edouard asked his father if he could visit one of his friends. Edouard informed his father that he would be taking the bus to the school. When he got to school, he went straight to the first class scheduled for the day. The second class was a physics test. He took a deep breath before opening the exam paper and told himself, "No more lies, no more hate, and no more anger." He opened the exam and began to answer the questions on it. After attempting the exam, he asked his teacher if it was possible to grade it right now. The teacher granted his request and graded it. After that, the teacher handed it over to Edouard. Edouard took the exam papers and walked out of the classroom. When the teacher found out that Edouard had a very high grade, she was shocked.

To get to Hassan, Edouard took a bus to Mike N Nie's. When the restaurant called Hassan to tell him that Edouard was waiting for him, Hassan was relieved. He got there as soon as he could. Hassan walked into the restaurant and went straight to Edouard, who was waiting for him at the table.

"Hi, Edouard. I came as soon as possible. Sorry for being late," Hassan apologized to Edouard for the wait.

"Hi, I am sorry to come here, but I didn't know your address or phone number. I figured if I came here, I could meet you," Edouard apologized to Hassan for coming to the restaurant.

"Is everything alright with you?" Hassan asked anxiously.

Edouard took the exam papers from his bag and gave them to Hassan. Hassan looked at his grade and smiled.

"You did great." Hassan greeted the kid while smiling.

"I am sorry if I offended you. I don't know who you are or what you want, but please don't tear us apart." Edouard burst out in tears while sharing his grief with Hassan.

"You didn't offend me and I have no intention to tear your family apart. I don't know how things will go between your father and me, but I want you to know I am on your side." Hassan acted calmly and tried to clear Edouard's confusion.

It was then that Hassan approached Edouard, wiped his tears, and asked him to smile. Edouard embraced Hassan and said, "Thank you!"

"I will take you home, and we can all have dinner together," said Hassan.

"Okay." Edouard nodded.

In a phone call with Klaus, Hassan asked that he wait until he arrived for dinner. Klaus was taken aback by Hassan's proposal. Hassan drove Edouard home and texted Klaus to open the door before he arrived. A smiling Hassan and Edouard walked past Klaus as he opened the door; he was left shocked at what he witnessed.

Hassan stopped and said, "Hey, Klaus. We are here."

"What is happening?" Klaus asked as he was confused by what he was seeing.

"Relax, Edouard wanted to see me. So, we talked," Hassan assured Klaus that everything was fine.

Klaus was handed over the test paper. He was pleasantly surprised. Then he grabbed Edouard by the arm and said (with tears), "I am very proud of you." and gave him a forehead kiss. Then, as Edouard left the room, Klaus held Hassan in his arms and kissed him. In Hassan's ear, Klaus whispered: "Thank you." Then Hassan went into the room with a smile and Klaus shut the door behind him.

***

A few months later, John was sitting in his office when he received a message from Carlos. John smiled as he saw the message and read it out loud. John began dating Carlos, but he was afraid to face Hassan. Every time Carlos came over, he made sure that Hassan wouldn't be able to see him. His goal was for Carlos to feel comfortable around Hassan and not be intimidated.

***

As Pete made his way to a restaurant, Pete's cousin spotted him. His cousin walked up to him and started a conversation with him.

"Hey, Pete!" Jonathan greeted.

"Jonathan! Oh, my God. How are you?" Pete was rather surprised.

"I am okay. Thanks. How are you doing?" Jonathan asked about Pete's health.

"I am doing great," Pete replied.

"Why didn't you come home?"

"This is home. I work at this restaurant."

"You look great."

"I have a job and an amazing boyfriend."

"I am happy for you."

"I have to go now; I have a job to do."

"See you around."

"See you."

To get to the restaurant, Pete took a brisk stroll. In the meantime, Jonathan picked up his phone and made a call to Pete's dad, and told him that he had located Pete.

****

On the other hand, there was a tense moment going on between Kyle and Cara. Kyle made a call to Cara. He wanted to tell her how much he cared for and respected her. It was as if her vines were tingling with electricity. She put the phone near her heart, took a deep breath, and smiled after the call ended.

***

Andrew was at the hospital at the time. During his walk around the hospital, he was feeling a bit down. Despite being surrounded by his roommates, his loneliness was growing. There were too many nonsense hookups and one-night stands one could have. He was looking for something consistent this time. He bought a sandwich from the cafeteria and sat down at a table alone. He took a good look around. He was the only one who was sitting by himself in the room. He walked to the restroom, leaving his sandwich on the table. He closed the stalls' door, put his hand on his mouth and cried without making a sound.

****

Hassan was at work when he heard a familiar voice, "We met again." Hassan turned around and he saw Roy.

"Hey, Roy, right?" Hassan tried to confirm his assumption.

"Yes, Roy. How are you doing?" Roy confirmed it.

"I am doing well. Thanks for asking, and yourself?" Hassan asked.

"I am doing great. I wished we had time to talk on that day. I would like to get to know you. What do you think of going on a date?" Roy instantly asked Hassan out for a date.

"A date!" Hassan smiled while replying.

"I want to know you. Here is my number (he gave his business card to Hassan). No pressure, when you are ready, give me a call." Roy left after handing his business card to Hassan.

Hassan took Roy's business card and said, "Okay." Camille saw their little interaction; she walked to Hassan and talked to him.

"Was that my brother Roy?" Camille pretended as if she was not aware of the situation.

"Yes." Hassan answered.

"What did he talk about?" Camille asked.

"He asked me on a date and gave me his number," Hassan told Camille about his conversation with Roy.

"Hassan, I hope you will understand. He is my brother; he is healthy and…"

"No need. Here is his card." Hassan interrupted Camille.

He gave Camille Roy's card and walked away. He felt he was burning inside. He went to his desk, collected his stuff, and left. When he reached home, Klaus was calling him.

"Hey"

"Hey, Hassan. Are you free tonight?"

"I can't. I have to do something."

"Tomorrow?"

"Can I ask you something?"

"Sure."

"Are we heading somewhere, or are we just having fun?" Hassan asked Klaus.

"We are having fun," Klaus replied.

"That's what I thought. Anyway, I am done with the fun. Now I am searching for something serious." Hassan informed Klaus about his relationship plans.

"So, this is goodbye?"

"Yes. Goodbye, Klaus."

"Goodbye, Hassan."

Hassan hung up the phone and went to his room, where he cried in the dark. Face tissues in hand, he sat in his bedside chair.

As Andrew walked into the house, he was sobbing. He went to Hassan's room when he saw Hassan's shoes.

He entered the room, "I need to talk to you." Andrew said. He saw Hassan on the floor, crying.

"What happened to you?" Andrew tried to inquire Hassan about his reason to cry.

"HIV happened to me. What happened to you?" Hassan raised his voice a little out of frustration while talking to Andrew.

"Loneliness happened to me."

"Sit here. Misery loves company."

"What are we going to do?"

"I have no idea."

"At least you have Klaus. I have no one."

"I ended it with him."

"Why?"

"He doesn't want anything serious," Hassan said with a sad smile.

"Now, we are miserable people who are cursed to be alone."

"I don't know. I don't know. I am tired. I will lay on the bed."

"I will lay next to you."

And just like that, they both fell asleep on the bed.

The next morning, everyone sat around the table for breakfast. Andrew hadn't cracked a joke or even said a word in the whole conversation. Hassan was using his laptop to take notes and jot them down on paper.

Cara looked at Hassan and Andrew and asked them, "What happened to you both?"

They replied together, "Nothing."

John looked at Hassan's laptop and then said, "Why are you searching for adoption agencies?"

"You said it, adoption agencies," Hassan said.

"What happened to you?" asked Cara.

"Leave him be. Not everyone can be happy all the time." Andrew intervened.

"You both need to talk now," Ethan said in a rather commanding tone.

"I want to adopt a baby. Is that too much to ask?" Hassan informed the group of his plans.

"What is going on?" Pete asked as he was confused about the whole situation.

"Simply, I am miserable, and he is lonely," Hassan said on behalf of Andrew.

"That is not simple." Kyle denied Hassan's statement.

"Yes, it is simple." Andrew corrected Kyle.

"What happened with you and Klaus?" John asked Hassan.

"Nothing happened. We are not together." Hassan told everyone about his breakup.

"Is 'that' because you broke up with Klaus?" Ethan asked, pointing to his laptop and referring to his adoption plan.

"He wasn't looking for something serious, and I knew that already. So, no, because of Klaus." Hassan said.

"What is it about?" John asked.

"I was called a whore last week, and people are afraid of me," Hassan replied.

"Why?" asked Pete.

"Because he is living with HIV." Andrew intervened again but yelled this time.

"I am sick of all the drama. All I want is a normal life. I no longer have the energy to stand up every time I fall," Hassan replied.

Hassan shut his laptop, collected all his stuff and made his way to a nearby restaurant. After Hassan and Andrew left the house, everybody was staring at each other in disbelief, wondering, "What just happened?"

***

Pete was on his way to the restaurant to begin his workday when he noticed someone walking toward him. He was stopped by a voice he recognized. Seeing his father, he turned to face him.

"What are you doing here?" Pete yelled.

Hassan saw Pete talking to a man angrily. He went outside and watched Pete.

"I just came to ask you for forgiveness." Pete's father replied.

"Forgiveness? For destroying my life or letting my mother die? For what are you asking for forgiveness?" Pete started venting out.

"You are my blood." Pete's father tried to convince his son.

"What kind of blood made you watch me get raped by a man over and over just because I am different?" Pete asked in anger.

"I was blind." Pete's father tried to justify his doings.

"You need to go. I can give you nothing," Pete told his father to leave.

"You are my son." Pete's father was adamant about convincing him.

"You are not my father. You are d…", Pete was about to lash out, but suddenly, Hassan interrupted him.

"That's enough. Pete, go inside." Hassan told Pete.

"Are you his partner?" Pete's father asked Hassan.

"I am not, but I am a family," Hassan replied.

"Please tell him I am sorry," Pete's father asked Hassan to convey his apologies to Pete.

"If you are sorry, you need to tell him yourself." Hassan denied the request.

"He won't listen to me," Pete's father expressed to Hassan that what he suggested was impossible.

"Not by word. What you have done is unforgivable. You need to show him that you are sorry." Hassan came up with a suggestion for Pete's father.

"I will. Thank you." Pete's father thanked Hassan and left.

Hassan entered the restaurant after Pete's father left. Upon entering, he found Pete sobbing in front of the front door. Hassan went up to him and hugged him; once he calmed down, the two walked together to the office.

"I am sorry all of that happened here." Pete apologized to Hassan for creating a scene. "No need to worry," Hassan replied. "Why did you stop me?" Pete asked. "Simply, he is still your father, and I could see the regret in his eyes. The kind of regret that eats you up alive. I stopped you because I don't want you to be like him. I don't want you to regret or do anything wrong." He assured Pete, then to distract him, he said, "You can take the day off. Go home and have some rest."

****

Hassan took Ruby to the hospital a few weeks after the accident. She was in poor health. In the waiting room, Hassan was waiting for news about Ruby and her baby. He put his head in the palm of his hands when he heard a familiar voice. It was Klaus who had tears in his eyes when Hassan looked at him. He sat down next to Hassan in the front row.

"What are you doing here?" Hassan asked Klaus.

"Charley is having her baby," Klaus replied while sobbing.

"Then, why are you crying?" Hassan asked.

"Her cervix did not dilate enough. She is having a very rough time." Klaus replied.

"I am sorry to hear that."

"What are you doing here?"

"Ruby is a girl who works with me; she is here having a baby." Hassan explained.

"I miss you."

"Klaus, please, not now."

"Can I ask you one last thing to do for me?"

"What is it?"

"Just talk to Charley."

"Okay, I will. Can you wait here for Ruby?"

"Sure."

"Where is Charley's room?"

"Third one on the left."

Hassan went to Charley's room; as soon as he entered, Charley looked at him and yelled, "Where the hell have you been?"

"I am here now," Hassan answered.

"You promised me you would be here for me," Charley kept on yelling.

"I am sorry." Hassan apologized to her.

"I can have the baby now. Can you please press this button?" Charley said.

A nurse and an obstetrician arrived as soon as he pressed the button. Hassan sat behind Charley and held her hand as he listened to her story. Even though she was exhausted, she continued to push until her baby was born. She was completely drained. Then Hassan checked to see if she was okay, so he could leave.

"I am not done with you." Charley tried to stop Hassan.

"I am afraid I've done my part," Hassan replied.

"You left him alone and miserable." Charley blamed Hassan for Klaus' mental health.

"He doesn't want something serious, and I can't be with him." Hassan justified his stance.

"He has changed, thanks to you." Charley told Hassan.

"I have to go." Hassan insisted on leaving.

"You are not going anywhere until you answer me." Charley ordered Hassan to stay.

"Do you want your father to be with me?"

"We all want that."

"What? Your father didn't tell you that I am living with HIV."

"Stop being the king of the drama. He is falling for you, and he doesn't care."

"I do."

"I know the feeling of someone hurting you so badly and you excluding everyone who wants to be close to you. He needs you, and you need him. I've never seen him smile the way you make him smile. So, stop this nonsense. Look me in the eyes and tell me you are better off without him, and I will let you go." In the midst of all this, Hassan was crying and Klaus had entered the room as well.

"I want to be with you." Klaus came in and confessed to Hassan.

Klaus moved toward Hassan, put his hand on his face, and wiped his tears.

"My family needs you. I need you." Klaus told Hassan.

"I no longer have the energy to stand up every time I fall. I am exhausted," Hassan replied.

"I know," Klaus said.

"Hassan, I want you to name my girl," Charley asked Hassan to name her newly born baby girl.

"What about Peyton?" Hassan asked.

"Peyton! I like it. Peyton, meet your grandfather and future grandfather." Charley liked Hassan's suggested name for her baby.

"I have to check on Ruby." Hassan tried to excuse himself to check on his colleague.

"She is waiting for you," Klaus said to Hassan.

Klaus stopped Hassan and kissed him on the cheek. Hassan just smiled and walked out of the door. He went to Ruby's room. She had a healthy baby boy, and he was happy to see it. He smiled as he held him between his arms. Ruby named her son William. He was ecstatic that she had found happiness. He walked out of Ruby's room, but he wasn't feeling well at the time. Eventually, his vision blurred as he leaned against the wall. He felt his heart stopped for a moment, and he passed out shortly afterward.

Hassan found himself walking on a road that seemed to have no end. Suddenly, a huge gate appeared in front of Hassan's eyes. He heard his father's voice, "What are you doing here? You need to go back. You are not finished. Go back and be happy."

Hassan felt a weight on his shoulders. A hospital bed was waiting for him, with Klaus standing next to him.

"What happened?" Hassan asked.

"You had very low blood sugar levels, so you fainted. I was very worried about you." Klaus told Hassan.

"Do you want to be with me?" Hassan asked Klaus if he was sure about his decision.

"More than anything," Klaus assured Hassan of his decision.

"You should know I am a drama queen," Hassan humorously warned Klaus.

"You just had a rough time." Klaus moved his hand over Hassan's face while replying to him.

"I've missed you." Hassan told Klaus.

"Me too." Klaus shared his mutual feeling with Hassan.

****

Just when there was a cherishing moment between Klaus and Hassan, he saw that Hassan's phone was ringing. It was John calling him.

"Hey, John." Klaus answered the phone.

"Who am I talking to?" John did not recognize Klaus' voice.

"It's me, Klaus." Klaus answered.

"Hey, Klaus, where is Hassan?" John asked.

"He is in the hospital," Klaus replied.

"Which Hospital? Is he alright?" John asked Klaus anxiously.

"Ruby and my daughter had a baby, that's why we are here. Nothing to worry about. Oh! And John, Hassan will spend the weekend with me." Klaus informed John about the situation.

"Okay. Thanks for letting me know. Have a good day." John became certain and bid greetings to Klaus.

"You as well." Klaus returned the courtesy.

John hung up and smiled. At the same time, Andrew entered the house, removed his shoes, and walked in the living room. When he looked up, he saw John smiling.

"Why are you smiling?" asked Andrew.

"Hassan is back with Klaus," John told Andrew.

"Oh, no," Andrew said as if it was not supposed to happen.

"Aren't you supposed to be happy for him?" John asked.

"We agreed to be together if we couldn't find anyone. Now my backup plan is gone." Andrew told John about his plan.

"You are a drama. There is someone I want you to meet. Come join us for dinner." John smirked and invited Andrew over to dinner with Julian.

"Okay." Andrew agreed.

"I want you to meet Julian. He is sitting next to Ethan." John introduced Andrew to Julian.

"You want me to meet glasses!" said Andrew.

"Don't be shallow. You will love him." John replied.

Andrew and John joined the others for dinner.

"Where is Hassan?" asked Cara.

"He will spend the weekend with Klaus," John announced Hassan's reunion with Klaus.

"Klaus!" Kyle was surprised.

"Yes, lucky Klaus," Andrew confirmed Kyle.

"Hassan, the only one I haven't met yet." Carlos joined the conversation.

"Believe me, your boyfriend doesn't want you to… ouch! (Talking to John) Did you just kick me?" Andrew had exposed John and everyone laughed.

"Yes, and I am willing to do it again if you don't keep your mouth shut," John warned Andrew by whispering in his ears.

"How dare you!" Andrew said in a fake British accent.

"Am I missing something?" Carlos was nascent about the whole situation.

"Your boyfriend is threatening me with kindness. Isn't that true, John?" Andrew taunted John.

"Why would he threaten you?" Carlos tried to get the gist of what was happening.

"If you want me to save your ass, bring me a real date, not glasses." Andrew challenged John by whispering in his ears.

"Deal," John whispered back to Andrew.

"I was about to ruin your boyfriend's surprise, so he kicked me and is now threatening me," Andrew diverted everyone's attention by changing the topic.

"What surprise?" Carlos asked.

"Oh, boy!" Andrew shrugged.

"It won't be a surprise if he tells you. Am I right?" Julian joined in.

"You are right, Julian." John approved.

"In that case, okay. I can wait." Carlos said smiling.

"So, Andrew. Is that your thing? Using humor as a shield to hide your feelings of loneliness?" Julian infiltrated Andrew's insecurities by asking him about his loneliness.

"How did you know?" Andrew was surprised.

"I am wearing glasses. I am not blind." Julian replied.

"I am sorry." Andrew apologized for his joke on Julian's glasses earlier.

"It's okay. I am used to people calling me glasses." Julian told Andrew that he was not offended by Andrew's remarks about him.

After dinner, they all sat together, talking. Andrew took Julian to talk privately.

"Julian, why are you here?" Andrew asked.

"I am here to meet you and get to know you. Since I saw you walking with John, I could not stop thinking about you. I am in love with you," Julian confessed his love to Andrew.

"What?" Andrew was shocked at his sudden confession.

"Relax, you are not the only one with humor," Julian replied.

"For a second, you freaked me out. But…" Andrew was interrupted by Julian while speaking.

"Just shut up and kiss me already," Julian asked for a long-awaited kiss.

"Oh, sassy." Andrew liked that.

A long kiss was exchanged between Andrew, who moved closer to Julian. Everyone was paying attention to what was going on. Ethan screamed at the top of his lungs.

"Yes, Julian. Show him who is the boss." Everyone looked at Ethan.

"What?" Ethan was taken aback by an awkward silence.

"At this moment, I couldn't recognize you." Pete said.

"Shut up, you love me," Ethan replied.

"True," Pete agreed.

"Don't you have anything else to gossip about?" Andrew tried to stall everyone.

"We will leave you alone." Cara agreed that they all should give Andrew and Julian some privacy.

***

To Hassan's surprise, Klaus took him to his home. He escorted him to his room so he could wash up. It was only fitting that Klaus gave Hassan some of Edouard's clothing after the shower. Klaus then put Hassan's clothes in the washing machine. Hassan joined María in the kitchen as she prepared dinner.

"You are back?" Edouard asked Hassan.

"Yes, I am," Hassan replied.

"I am happy to see you again," Edouard shared his happy feelings with Hassan.

"I missed you," Adrien was also happy about Hassan coming back home.

"I missed you too," Hassan exchanged mutual feelings.

Once Hassan was back, Klaus was relieved to see him. He believed that the days ahead would be filled with joy and happiness. However, even though he was not mistaken, something was wrong in his mind. When things went wrong, it was going to be painful. But he did not know why he had that feeling. He told himself, "My family is here. He is here. What wrong could happen now?'

After having a rough day, Klaus put Hassan to bed after dinner. They both undressed and he laid Hassan down on the bed. Under the covers, his arms were wrapped around Hassan's waist, and they kissed. When Hassan fell asleep, he did so between Klaus's arms, where he felt safe. As soon as he was sure Hassan was asleep, Klaus closed his eyes and tried to fall asleep. A vision of himself walking through the woods appeared to Klaus after he closed his eyes. Nothing but trees could be seen in the distance. In an attempt to find a way out of the woods, he kept walking. To keep walking straight, he heard the voice of his wife, who said, "Turn left." He turned to the left and continued walking until he found Lizzie standing on the edge of a cliff wearing all white.

***

Chapter 17: Broken Hearts Begin to Mend

"It's a beautiful thing to meet someone who makes you forget all your troubles" -
Anonymous

"I need you. We all need you," said Klaus to Lizzie as he moved in closer to hug her and tell her that how much he missed her.

"I know. You finally found my gift for you," Lizzie replied.

"Which gift?" Klaus was surprised.

"I am watching over you and our kids. I looked everywhere; I couldn't find someone to repair my family. Then I found him." Lizzie told Klaus.

"You found whom?"

"The one who is sleeping now between your arms."

"You are talking about Hassan?"

"Yes, Hassan. He is a warrior. I know that you like him a lot more than you realize."

"I do like him a lot," Klaus confessed.

"You have your next steps mapped out, but you need to promise me one thing."

"Anything you want."

"He brought my family together, so promise me to not give up on him."

"I won't give up on him, but can you tell me why?"

"You had the feeling that something will go wrong. Am I right?"

"I might be mistaken."

"You are not mistaken. A dark cloud will come and will take him to a very dark place, and you are the only one who can help him. It might seem that he no longer has the energy to stand up, and he will be exhausted. The dark cloud will consume him from the inside. But he is a worrier, and he will rise again."

"What could happen?" Klaus was now worried.

"Look down the cliff, and see for yourself."

Klaus looked down the cliff. He saw Hassan chained in every inch of his body floating in the air. Klaus cried and asked, "What is happening to him?"

"It's the dark cloud that will drag him to a very dark place."

Suddenly, the chain on Hassan's neck pulled him down to the cliff into the water. Klaus looked, cried, and screamed for assistance, but it was too late. Hassan has drowned in the water already. Klaus suddenly awoke out of his sleep and immediately kissed Hassan's forehead. Klaus' tears began to fall out of his eyes, and said, "Please fight for me, fight for us." At that moment, Klaus felt something he hadn't felt for years. On his body, he felt Lizzie's hand.

"I know you are here. Thank you." Klaus said.

The next morning, Hassan woke up early; he kissed Klaus, then walked with Bella to the kitchen to prepare breakfast. Then, Hassan went to everybody's room to wake them up. Klaus was the last to rise. That, too, when Hassan began to jump on the bed, and Klaus had to obliged. Klaus brought down Hassan and kissed him. He breathed deeply and said, "You are here."

"Yes, I am here," Hassan replied.

"I've missed your smile. I've missed the way you always wake me up in the morning. You know I am still naked," Klaus said to Hassan and winked.

"I hate to break it to you, but now you are not getting any. They are waiting for us downstairs." Hassan turned down Klaus' advances.

"Daddy is hungry," Klaus said, biting his lip.

"If daddy could wait a little bit longer, then daddy can eat as much as he wants to." Hassan tried to stall Klaus.

"Daddy is okay with that. You have no idea how hungry daddy is."

Hassan looked at Klaus and smiled. Klaus melted in that smile. He quickly got dressed and joined everybody downstairs for breakfast.

Andrew woke up next to Julian. He glanced under the blanket to see if he was wearing underwear.

"Relax, nothing happened," Julian assured Andrew.

"I remember nothing. Can you help me with some details?" Andrew asked Julian to clear his confusion.

"I am still wearing my clothes. What kind of a person do you think I am?" Julian was taken aback.

"The kind who sleeps with clothes next to a drunk naked hot guy like me," Andrew replied.

"Get over yourself. I will never take advantage of anyone," Julian said.

"Why didn't you take advantage? I tried to have sex with you, and you stopped me. That is embarrassing." Andrew disclosed his desire subtly.

"I put you in the bed, and we talked after you danced naked," Julian reminded Andrew of his shenanigans.

"Did anyone see me dancing naked?" Andrew asked.

"Only me. Look, Andrew. You are an attractive guy, but I am not that kind." Julian was consistently trying to sway from Andrew's verbal sexual advances.

"I am sorry if I hurt you." Andrew apologized to Julian for his behavior.

"You didn't. You kept talking about how lonely you are and your relationship with Hassan. It seems everyone in this house is somehow connected to Hassan," Julian said.

"Well, we're all here because of him," Andrew said.

"I want to know him, but first, I want to know you." Julian showed his intention to get acquainted with Andrew first and then Hassan.

"Me?" Andrew was shaking.

"You are smart, sexy, and a nice person. So, are you interested in glasses?" Julian asked Andrew if he was interested in taking things further.

"I am sorry that I offended you, but I would like to get to know you," Andrew said.

"Good."

"Why do you want to know Hassan?" Andrew asked Julian out of curiosity.

"I am a psychiatrist, and I am always curious about people. What I have heard from John made me more curious about the people who live in this house. You and Hassan are very similar; both are smart, emotional, and have a unique quality," Julian opened up about himself and his opinions on Hassan.

"So, you are not interested in me. You are just curious, nothing more." Andrew showed his reservations.

"I asked John to introduce me to you not because I was curious." Julian tried to justify his stance.

"Why do you want to know me?"

"I told you, you are unique. And I want to be someone with your quality to rock my world. So, are you up for it?"

"Before you try to date me, you should know a few things. I used to be easily impressed and easily won. But now I'm a different person; having me in your life is going to take work. I need someone who will make a genuine effort to know me, care for me, and continue to strive to make me happy, just as I would for you. If you want me in your life, you're going to need to earn me." Andrew said in a very serious tone.

"I saw you naked!" exclaimed Julian.

"Enjoy the view as much you can." Andrew was back to being himself.

"You are interesting, and I would like to know you better."

***

After breakfast, Klaus wanted to talk to Hassan, so he took him to his room.

"Can we talk about us?" Klaus asked.

"Sure, we can," Hassan replied.

"What are we now?" Klaus tried to clear the air between him and Hassan.

"We are two people who find each other attractive so we're dating," Hassan replied.

"So, we are dating now!" Klaus assumed.

"Use your words," Hassan said.

"I don't want to lose you again. So, I am okay with dating for now and a relationship later. But what I need to know now, why did you break up with John?" Klaus couldn't hold back.

"Simply, he lied and hid things from me."

"I would never lie to you." He assured Hassan.

"You already did." Hassan was quick to point out.

"When?"

"At the hospital. When you said Charley's cervix did not dilate enough. She is having a very rough time."

"She came up with this idea with the nurse so she would be able to talk to get back to me," Klaus explained himself.

"Ah!"

"I am not going to lie. I like you so much, and my kids love you." Klaus sighed and continued, "I am very serious about you."

"When I asked you if we are just having fun or we are heading somewhere, and you said, 'we are having fun.' I felt I was not good enough for you or for anyone." Hassan started crying at this point.

"That is not true. After you, I slept with many guys to prove that you were just another meaningless one-night stand. Then I realized I was into you, but it was too late."

"So, here we are."

"Yes, here we are. I met you, and you became my best and dearest friend. There was always a spark between us. I will never let you go. Please tell me you won't either."

"I want to be with you."

Klaus gave Hassan a passionate kiss and said, "Change, we have to go to the hospital to bring Charley and her baby back home.

"I need to check on Ruby also," Hassan said to Klaus.

"Okay," Klaus replied.

"Also, I need to go home and change," Hassan remembered he hadn't brought anything over with him.

"Bring some with you and leave it here. Don't forget the sexy underwear."

"Anything else?" Hassan asked with a smile.

"That's all," Klaus replied.

Hassan and Klaus were going to the hospital, and Hassan went home from there. Cara, Pete, Kyle, Ethan, John were all hanging out in the living room.

"Andrew, why don't you call Julian and ask him to join us for dinner," Cara asked.

"He has some stuff to do," Andrew replied.

"John, what about you, Carlos?" Cara inquired John about Carlos.

"He won't come again," John responded.

"What do you mean?" Ethan asked.

"Did you kill him or something?" Andrew asked out of intrigue.

"We broke up last night." John disclosed the news to everyone.

"And you are telling us now?" Pete raised his tone a little.

"Tell us what happened?" Kyle asked.

"We were having sex, and I said something that made him leave." John shared the saddening story while blaming himself for the breakup.

"What did you say?" Andrew asked.

"I called him 'shorty'," John said.

"You said what?" everyone asked altogether.

"I think it's for the best," John said.

"If Hassan finds out, he will kill you," Cara warned John.

"Why is everyone afraid of him? He is tiny." Andrew tried to ask why everyone feared Hassan.

John felt something pull his ear hard, "Ay, Ay." John shrieked in pain. Everyone looked at John and saw that Hassan was holding John's ear between his fingers. Everyone jumped.

"You did what?" Hassan asked in anger.

"What did you hear?" John tried to make sure how much Hassan had gotten to know.

"Everything. Now, tell me, why didn't you talk to him?" Hassan inquired John angrily.

"Please, just let me go," John screamed.

Hassan released John, and he put his hand on his ear, which was hurting.

"He wanted me to leave after I called him what I used to call you," John told Hassan.

"And you let him go without saying anything?" Hassan asked.

"Yes," John replied.

"I don't believe you. For more than six years, you were excellent at making up excuses. What happened to you?"

"I didn't want to lie to him or give him excuses. I did it all with you, and I lost you." John sighed.

"Great, but now you lost him too." Hassan reminded him.

"It was never meant to be."

"Give me your phone."

"What?"

"Give your phone, or give me his number."

"It's over."

"Your phone," Hassan yelled at John.

He silently gave Hassan his phone. Hassan went through his contacts and found Carlos' number. He dialed it, and Carlos answered it at the first ring.

"May I speak to Carlos?" Hassan asked for Carlos.

"Speaking," Carlos replied.

"I am Hassan, John's ex. Are you available to meet up and talk?" Hassan asked.

"Didn't John tell you we broke up?" Carlos asked Hassan.

"We need to talk." Hassan maintained a serious tone as if he meant business.

"About what?" Carlos grinned a little.

"About you."

"Okay, I am free now."

"Okay, come to my house."

"But John will be there."

"I am waiting for you."

Carlos ended the call. Hassan looked at John and said, "Don't try to say or do anything."

Andrew whispered in Cara's ear, "Now I can see why everyone is afraid of him."

"How is Klaus?" Pete asked Hassan.

"He is good," Hassan replied.

"Details, please." Cara wanted to know a little more from Hassan.

"Are you a couple now?" Andrew joined in too.

"We are dating, not a big deal," Hassan replied.

"It is a big deal," Kyle told Hassan that it was something not to be taken lightly.

"Look at your face. You don't have that sour puss you had for weeks," Andrew teased Hassan.

"I don't know what you are talking about?" Hassan showed that he was confused.

"Welcome back." Ethan jumped into the conversation.

Hassan smiled and sat between them and talked. After a while, the doorbell rang and Hassan got up from his seat to open the door.

"Hi, I am Carlos," Carlos greeted Hassan.

"Hi Carlos, come in." Hassan welcomed him in.

"So, you are Hassan," said Carlos.

"Finally, we met. I've heard so much about you." Hassan tried to tell Carlos that he was curious to meet Carlos.

"Really!" Carlos was surprised.

"Your boyfriend cannot stop talking about you," Hassan told Carlos.

"John is not my boyfriend, not anymore." Carlos corrected Hassan.

"May I ask why?" Hassan asked.

"He didn't tell you?"

"Maybe I want to hear from you?

"He is still in love with you, and I cannot compete with you," Carlos sighed.

Hassan tried to mentor Carlos at this point by saying, "With every new relationship comes struggle and old baggage. Save yourself the heartache and realize that these people are likely going to be in your boyfriends' life for as long as you are, and maybe longer. When it comes to exes, you have to surrender because you cannot compete with them. There will always be too much history with the exes, so don't put yourself in this situation. I understand that you are intimidated by me, and that is not your fault. That's on him because he was afraid to introduce you to me. I can understand why, but I cannot understand why you both are not together? Is it because he said something?"

"But he is not over you yet," Carlos was beating about the bush.

"John would have never gotten involved with you if he wasn't over me. I've been with him for six years; I know this much." Hassan assured Carlos.

"But he cheated on you, and that's why you ended it with him," Carlos insisted.

"Yes, John cheated on me only because he thought I cheated on him. I ended it with John because he lied to me and kept things hidden from me, which will never happen with you." Hassan consistently kept on justifying John's actions.

"Why?" Carlos asked.

"He has learned his lesson. That's why he didn't tell you anything after you stormed out. Give him a chance because you are into him, and you worth knowing," Hassan said to Carlos.

"Do you think I am worth knowing?" Carlos asked.

"Everyone in this house knows that. John knows it. Look, I got it. You didn't know what to do in a situation like this, but you are smart enough to understand that you need to accept him as he is and add your value to build a strong relationship with him," Hassan replied.

Carlos had been feeling helpless, but Hassan helped him out and finally cleared the rifts between him and John.

"I don't know what I have to do?" confused Carlos said.

"No one knows, but for a start, go talk to your boyfriend," Hassan replied.

"Okay," said Carlos in a low tone.

"Just look behind you." Hassan redirected Carlos toward John.

Carlos turned and saw John standing behind him. He looked at him. John moved to Carlos, placed his head on his forehead, closed his eyes, and said, "I am sorry that I hurt you."

Carlos threw himself into John's arms. Hassan saw them, smiled at them, and said, "Don't do it again, or I will kick your ass from here on."

John kissed Carlos's head and told him, "He will do it, believe me."

Carlos laughed, and John looked at his face and said, "I want to see your smile."

Andrew walked to Hassan and hugged him, "I missed you."

"I was only gone for 24 hours," Hassan said as if it was not a big deal.

"Don't play dumb. You know what I am talking about," Andrew objected.

"I know," Hassan nodded.

***

When Brian received the text asking him to be more patient, Brian was sitting there, and he was planning on something. Brian went to the cellar where he is keeping the painting. The painting had a chain. He approached and touched it. He closed his eyes and recalled Hassan's naked body in his bedroom when he touched it. His eyes were opened, he touched, breathed on the painting, and said, "It's just a matter of time." He laughed then.

***

Andrew went to Lake Ontario Park. He was sitting at the bank waiting to see Nie. Andrew looked at Nie and smiled when he appeared.

"You were right; they are back together. Now, what next?" Andrew said.

"Everything will be fine until one day, and everything will be changed forever." Nie made a vague statement.

"I met someone. I might end up with him," Andrew told Nie.

"You will be with someone, but I don't believe you will end up with this one," Nie said with a smile of a wise man.

"Oh, no. What is wrong with me?" Andrew got confused suddenly.

"Everything has the perfect time for it, and your time hasn't come yet," Nie assured Andrew.

"I don't want to be alone anymore." Andrew talked about his unwillingness to embrace solitude.

"What is waiting for you is worth being patient." Nie insisted Andrew wait for the right moment and the right person.

"What is waiting for me?" Andrew asked.

"You will be a wonderful husband and father, one day," Nie replied.

"I will be a dad?" Andrew smiled.

"You will be," Nie assured Andrew.

***

Hassan came over to Klaus' house. María opened the door.

"Welcome back." María welcomed Hassan into the house.

Hassan kissed María's forehead and said, "I am happy to see you every week as I used to."

"Did you bring some clothes?" María asked.

"Yes, I did," Hassan replied.

"That is a good sign." She thought of it as a good omen.

"Are they here yet?" Hassan asked about everyone's whereabouts in the house.

"They are upstairs," María answered.

"I will go and say hi to Charley and Peyton," Hassan said.

Hassan got upstairs and knocked at the door, "Come in," said Klaus. Then, with a big smile on his face, Hassan entered the room.

"Hi, everyone. Charley, how do you feel?" Hassan greeted everybody in the room.

"Tired," she replied.

"You need to take some rest. I have something for you and our little princess," Hassan advised Charley.

"Can I open it?" Charley asked.

"Sure," Hassan asked Charley to go ahead in opening the present.

Charley opened her present, she looked at it, then said, "You remembered."

"I have had it for weeks," Hassan replied.

"Thank you," Charley said.

"Can I see it?" Adrien asked.

"I would like to keep it to myself, but you can open Peyton's presents," Charley said to Adrien.

Adrien opened Peyton's presents. Baby clothes, toys, and baby shoes were all wrapped nicely. However, with Hassan's presents, Charley was happy.

"Is that your bag?" Edouard asked Hassan.

"Yes, Hassan will come every week as he used to. "Klaus replied to Edouard.

"Did you do what I asked you?" Edouard asked Hassan again.

"It is in this bag," Hassan replied.

"Can I see it?" Edouard asked.

"Sure," said Hassan.

He opened the bag and disclosed Edouard's and his mother's painting. Everyone looked at the painting. Edouard then smiled and hugged Hassan and said, "Thank you."

"That's nice of you," Klaus said to Hassan.

"Where is my gift?" Adrien intervened and asked Hassan for his gift.

"It's with me, but it is a secret," Hassan winked at him.

"You don't need to do that, but what did you bring for me?" Klaus jumped in and ask for his gift also.

Hassan whispered in Klaus's ear, "You should check my underwear." Klaus bit his lip and kissed Hassan. Then he held Hassan's hand and said, "See you later, guys. We have something important to discuss."

Then Klaus dragged Hassan and his bag to his room and shut the door behind him. Klaus moved closer to Hassan, he touched him and grabbed his ass. With soft kisses, Klaus undressed Hassan. He watched the blue underwear that Hassan was wearing and bit his lips. He said, "Daddy needs to eat." Hassan opened and unbuttoned Klaus' shirt. Klaus had his underwear taken off and held Hassan between his arms, putting him on the bed. Klaus watched Hassan and took his clothes.

"I want to breathe your skin," said Klaus. Klaus was like a hungry, thirsty lion wanting to hit his prey. He laid on his side, wrapped himself up, and kissed Hassan's neck. Hassan moved his hands all over Klaus' body. Then Klaus took Hassan's leg and put it around his waist while they kissed and moved it back to his butt. Hassan flipped on the bed to stand on top. He looked at Hassan, then put his hands on him and gave deep passionate kisses. The more they kissed, the more they fused. Klaus kissed Hassan's neck, and his tongue was shifting on it. Then he went down to his chest and belly. Hassan placed Klaus' head between the legs, held his hands tight, and his body crushed. Until Hassan finished, Klaus kept doing what he was doing. They kissed, and Klaus struck it, kissed his neck, and moved his tongue on Hassan's back. He grabbed his butt cheeks, bit them, and put his tongue on them. Then Klaus held Hassan's waist and moved inside of him.

Hassan moaned, and Klaus enjoyed that sweet sound. He then Klaus stopped and flipped Hassan to the top. Hassan moved up and down continuously. Hassan's vision of moving like that made the hungry lion crave more. Klaus stopped and brought Hassan toward him. Then Klaus moved Hassan to his bedside. He put him in a position to kiss him and moved within him, holding Hassan from his chest and leaving Hassan's leg with his hand. Hassan was moaning, which made him push more deeply. He continued to move until they finished together. He held Hassan between his arms and felt the chill on the skin. Hassan looked at Klaus' face and kissed him. They realized that they belonged together because of the strong chemistry and the flow of emotions, and they were comfortable expressing their feelings for each other. They were reaching the highest levels of intimacy after kissing. All Klaus thought about while watching Hassan was, "Am I in love?"

At the same time, Hassan was thinking, "What am I feeling? This feeling is so very different. Am I falling in love?" Hassan and Klaus took a shower together. Klaus took a photo of them together after the shower. Hassan asked Klaus, "Why did you take a picture?" Klaus looked at Hassan, smiled, and said, "I will take a picture of every day we spend together. I want to remember every day with you from now on." Hassan smiled at his answer, and Klaus hugged him. Hassan then placed his bag on that bed.

"What are you doing? We will sleep naked as usual after changing the sheets." Klaus asked Hassan.

Hassan opened his bag and took a small bag from it, and gave it to Klaus.

"What is this?" Klaus asked surprisingly.

"I know your birthday is tomorrow. Happy birthday." Hassan wished Klaus.

"May I?" Klaus asked, followed by a long sigh.

"Sure," said Hassan.

Klaus opened it, and there was a black wristwatch wrapped beautifully. Hassan looked at Klaus as he opened his gift.

"I love it. Thank you," said Klaus. He wrapped his hands and kissed Hassan. They laid down on the bed after changing the sheets.

"What are you afraid of?" Klaus asked Hassan.

"I am afraid of water because I cannot swim, heights which I discovered recently. I am afraid of sharp things, and I am afraid to be chained," Hassan replied.

"That is a lot of information, but I meant in life." Klaus rephrased his concern.

"Of being alone. What about you?" Hassan replied and asked Klaus the same question.

"To lose my kids, thanks to you, that would never happen. Can I ask you something?" Klaus asked him.

"Sure," Hassan replied.

"How did you know what I have to do precisely with my kids? Even when you had known them and me for one day?" Klaus asked.

"I followed my instincts," Hassan answered simply.

"Do you know what I love the most about you?" Klaus asked.

"No, what is it?" Hassan asked back.

"Your smile takes my breath away. What about you? What do you like the most about me?" Klaus confessed and asked what Hassan felt about him.

"You always capture my breath and my body yearns for you in a way I cannot resist, so I like everything about you." Hassan continued after a slight pause while locking his eyes with Klaus', "I am falling for you, and that scares me to death."

"Why does that make you scared?"

"I don't want to get hurt again."

"I would never hurt you," Klaus promised Hassan after a long sigh.

"I told you I am a drama queen." Hassan tried to lighten the mood.

"Don't ever say that again. And I want you to know that you belong to me."

They closed their eyes and kissed.

***

Carlos and John were in bed together. Carlos asked John, "Why didn't you want me to meet Hassan?"

John took a long sigh, "I was afraid that you would compare yourself with him. But I realized that I needed to make sure that I won't be blinded again," John said.

"I did compare myself to him, but he is right. I have no experience," Carlos said to John.

John laughed then said, "That is funny." Carlos looked at John, "What do you mean?" Carlos said.

"Hassan has been with only three guys, and he is talking about the experience! He surely is a piece of art," John replied to Carlos while laughing.

"He is. Now I see why you were afraid of me meeting him," Carlos said.

"I am sorry I made you feel like you were a replacement, you are not. And also, you cannot run every time we have a problem. You need to stay, communicate, and solve the problem together. If you want me, you need to fight for me, and I will fight furiously for you." John told Carlos that he would have his back no matter what.

"You would?" Carlos asked.

"I am scared to death," John replied.

"Why?"

"To be honest, I am thirty-nine years old who is dating twenty-one years old. What if you find a younger and healthier guy?"

"Would you fight furiously for me?"

"I would if you give me a reason to."

"You have it."

***

When Hassan received a text from his friend Oliver, he was in bed with Klaus talking, touching, kissing, and melting in each other's' arms.

"Monday, same time at the same place?" Hassan texted him back, "I will be there." Klaus read the text, and he asked Hassan, "Who is that?"

"This is Oliver, one of my friends," Hassan replied.

"I've never heard about him!" Klaus raised his suspicion.

"Do you know all of my friends?" Hassan asked rather sarcastically.

"No, but do I need to be worried about someone taking you away from me?" Klaus demanded reassurance.

"You are adorable, but you don't need to be worried," Hassan assured Klaus.

"Why shouldn't I?" Klaus asked.

"When I am with someone, I cannot even think about being with another guy," Hassan replied.

"In other words, you are mine," Klaus said.

"I might be." Hassan nodded.

Klaus smiled and kissed Hassan.

***

On a pillow, Oliver put his head and under his head, his right hand. He listened to "Tick Tack" on his hand watch. and fell asleep. He opened his eyes and found himself caught up in a chair, and saw Manauris wearing a hat with a pocket watch. His face couldn't be seen as his hat usually covered it. "Let me go," shouted Oliver. Manauris removed the watch from its repository sideways. The watch burned with fire suddenly. Oliver just thought of escaping because he didn't want to live in this hell all over again. Oliver shut his eyes, wishing the nightmare would be over. His eyes opened, but nothing changed. He collapsed and wept.

Oliver said, "No, not again."

But nobody could hear him as nobody was around, he screamed for help. Manauris threw a lighter out of his hand after he lit it up. The fire began to burn in high flames. He was not able to move; flames had now surrounded Oliver. He looked around; he could see nothing but fire.

"That's it." Oliver said to himself, "I will die here."

His body began to burn, and his veins could feel the boiling heat. He yelled for help, but no one could hear him. Again, he had to feel this unbearable pain all on his own. He screamed, "No. I can't... I can't. This is too much. All I wanted was to live without my biggest fears. And awfully wake up. He hung, and his eyes dropped with tears. I want to live a normal life." Oliver said. He laid the pillow on his head and shouted.

***

Hassan was working with a smile on his face on Monday. Tristan noticed him smiling and came over to say Hello.

"Someone's in a good mood," Tristan said.

"Tristan, how are you doing?" Hassan greeted him cheerfully.

"I am doing good, thanks for asking. How are you doing?" Tristan asked.

"I am doing great."

"You look different!"

"Yeah, I am happy."

"I can see that but cannot figure out why?"

"I spent my weekend with someone very special to me," Hassan replied with a smile on his face.

"I thought at least you would think of me, but it is okay. I just want to see you happy." Tristan sighed.

"If I hadn't met Klaus, you would be the first one on my list." Hassan tried to make Tristan feel better.

"You are just saying that not to make me feel bad, right?"

"I would never do that for you. You are very supportive and a real gentleman."

"You think of me as a real gentleman?"

"Stop it. You are a real gentleman. I am sure you will find someone."

"You know you are my favorite, friend, lucky Klaus. You owe me details." Tristan said and winked at Hassan.

"We can have lunch together." Said Hassan.

"Looking forward to it." Tristan approved.

***

Klaus had a detailed meeting with an entrepreneur. He went with his partner Nina. Robert was flirting with Klaus, but he was focused on the job. They agreed on many points and after hours of meeting with Robert and his team, Klaus and Nina left the office.

"Robert was a little flirty with you, and you didn't even look at him like you used to. Am I missing something?" Nina teased Klaus.

"I am seeing someone," Klaus replied.

"Klaus is seeing someone? Since when?" Nina was surprised.

"His name is Hassan, and he is very sweet," Klaus told Nina about his relationship with Hassan.

"When can I meet him?"

"Soon." Klaus gave her a vague answer.

"Okay!"

Hassan came back from lunch break, and he saw Klaus waiting close to the parking lifts. Hassan and Tristan walked to Klaus and said, "Klaus! What are you doing here?"

"Hey, I am here for a meeting. What are you doing here?" Klaus was stunned to see Hassan and Tristan there.

"I work in this building," said Hassan.

Tristan (cleared his throat) and Hassan said, "This is my friend Tristan. Tristan, this is Klaus and…"

"Nice to meet you, Klaus." Tristan interrupted.

"Hi, I am Nina, and you?" Nina introduced herself to Hassan.

"I am Hassan," said Hassan.

"Hassan, nice to meet you, and nice to meet you, Tristan." Nina greeted Hassan and Tristan.

"Nice to meet you, Nina." Hassan and Tristan shared mutual feelings with Nina.

"We are heading to grab a bite. Would you like to join us?" Nina asked.

"We already did, thank you," Tristan replied.

"Talk to you later." Klaus smiled and excused himself.

"Talk to you later," said Hassan.

After Klaus had given Hassan a kiss, he and Tristan left. Nina asked Klaus, "Is this the Hassan? What a coincidence to run into him like this!"

"Nina, can't you be happy for me?" Klaus sighed.

"Oh, my God. You are so into him. Is he something serious?" Nina asked.

"He is not something. He is a wonderful man." Klaus clarified.

"Klaus, the one who didn't want any relationship with anyone, is falling for him? What is special about him?" Nina asked.

"I don't know." Klaus was rather reserved in opening up. "I had no idea he worked in this building," he tried to change the topic.

"Interesting, but he seems very sweet. You couldn't take your eyes off him, and when he walked, you checked his ass. Tell me, what's his secret?" Nina was inclined to know more about Hassan. Klaus laughed off at her question.

"Finally, you found someone. I am happy for you," she said happily.

"Thanks, Nina."

***

Hassan went to meet Oliver in a park near Oliver's place after work. When Hassan arrived, Oliver was sitting on a bench, crying. He looked over at Hassan and hugged him.

"I had the same nightmare," Oliver told Hassan with tears in his eyes.

"No, not again. You did great work on yourself. We are so close." Hassan motivated Oliver.

"Please, take me home with you. I cannot sleep alone, not tonight at least." Oliver requested Hassan.

"I've got you." Hassan agreed.

"When will this nightmare be over?"

"You need to face your fears, only then will this nightmare be over."

"He did this to me."

"I know, but he is gone."

"Yes, he is gone, but what he did to me is still affecting me. Like a ghost haunting me."

"Oliver, we've come this far. Please don't let anything hold you back."

"I am trying."

"Stand up and come with me?"

"Where?"

"To my house."

"Okay."

Hassan took Oliver to his home and introduced him to everyone.

"I heard about you, but we never had a chance to meet." John went forward and welcomed him.

"It is nice to meet you all." Oliver shared his joy to meet everyone.

"Oliver, you need to shower to refresh, and we will talk afterward," Hassan said.

"Okay," said Oliver.

Hassan took Oliver to his room and showed him to the bathroom.

"I will bring some clothes to change from Andrew. You're both the same size, I think," said Hassan.

Hassan went downstairs and requested Andrew for some of his clothes to give to Oliver. Andrew went to his room, brought his clothes, and gave them to Hassan.

"One more thing," said Hassan.

"What is it?" Pete asked.

"No one should know about Oliver," Hassan warned everyone.

"Don't worry, we won't mention him to anyone," Ethan assured Hassan.

"Even to your therapist," Hassan said to John.

"Okay." John nodded.

So, after the shower, Hassan went to his room to put the clothes on his bed for Oliver to change. Hassan sat down with others in the living room. They spoke as they used to, about their day.

"Hassan, can I ask for a favor?" John asked.

"Sure, if I can," said Hassan.

"You can. Can you hire Carlos in your restaurant?" John asked Hassan for Carlos' job.

"What happened with his job?" Hassan asked.

"They are not paying him enough, and I want to support him, but he always refuses my help," John stated the situation to Hassan.

"Send me his resume," Hassan asked John.

"Can you ask him?" John asked in return.

"Ok, I will," replied Hassan.

"We are all wondering, who is Oliver?" Andrew was rather interested in knowing about Oliver.

"He is a friend of mine," Hassan answered.

"A secret friend!" Andrew smirked.

"Yes, a secret friend," Hassan said with no expressions on his face.

"You are full of surprises, aren't you?" Andrew exclaimed.

"You have no idea," John joined the conversation.

"What do you mean?" Pete asked.

"Every week, he was planning to do some pranks. Once, he scared me to death," Cara said.

"Look at him; he is an angel," said Kyle.

"Angel, my ass. He is evil," said John.

"That's true." Ethan agreed.

"I don't know what you are talking about!" Hassan was confused about what was happening.

"Really!" Cara said.

"It is time to pay back," John said to Hassan.

John held Hassan between his arms and asked Ethan to hold his legs. They placed him on the floor. Cara asked the others to tickle him everywhere. Hassan was held by Cara, Pete, Andrew, and Kyle, who all began tickling him. Hassan laughed hard and asked everyone to stop. Hassan asked them to stop, but John requested the magic word.

"No way." Hassan denied the request.

"Then you deserve it." John continued tickling.

"Are you okay?" Oliver did not know what everyone was doing.

"That's enough for today. We will do it again tomorrow," John said.

They left Hassan on the floor, holding his belly as he laughed a lot.

"That was fun," said Kyle.

"Oh, my God. Look at his face." Cara laughed.

"He is sored," said Pete.

"This is for scaring us a few days ago," Andrew directed their actions as friendly revenge.

"Oliver, can you give me a hand," Hassan asked Oliver for help.

Oliver helped Hassan stand up. He looked at everyone panting, "Traitors!" Hassan said.

"What can you do?" John challenged Hassan.

"You have no idea," Hassan said with an evil smile.

"Oh, no! He is back," Cara said.

"What do you think? Should we play truth or dare?" Andrew asked.

"Sounds good." Pete joined in.

"But please, no one ask for kissing except for couples." Hassan dictated the rules.

"You got it." Andrew agreed.

"Oliver, come and join us." Kyle invited Oliver to play.

They all gathered in a circle. Andrew brought an empty bottle and spun it. The first one was Hassan.

"Truth or dare?" asked Andrew.

"Dare," Hassan replied.

"I dare you to twerk." Andrew gave a challenge to Hassan.

"What? I cannot do that. I don't even know how to do it." Hassan was not up for the challenge.

"You know the rules." Andrew kept the challenge intact.

"Fine, I will try." Finally, Hassan surrendered and performed the task.

Hassan tried to twerk, and everyone laughed at him.

"Happy??!" Hassan said.

"That was fun." John teased Hassan.

Hassan spun the bottle and pointed to Andrew.

"Oh, boy." Andrew got scared.

"Dare or dare?" Hassan asked.

"You evil man," Andrew said.

"I dare you to do belly dancing." Hassan saw an opportunity to settle the score.

"You are mean." Andrew was not happy with the dare.

Hassan's eyebrows moved funnily. Andrew got up and began belly dancing. He was hilarious about how he was moving his leg. Everyone laughed, even Oliver. Andrew spun the bottle, pointing to Pete. Andrew looked at Pete and tried to find something.

"Can I dare you?" Andrew asked.

"Okay." Pete approved.

"I dare you to give your boyfriend a lap dance," said Andrew.

"No problem." Pete was rather happy performing the dare.

Pete asked Ethan to sit on the couch, and the dance began. Everyone was stunned by Pete's lap dancing skills.

"That was so great." Cara appreciated Pete's dance moves.

"Oh, my God. That was amazing," Andrew was surprised by Pete's moves.

"That is my boyfriend," Ethan said, biting his lip.

Pete spun the bottle and pointed to Oliver.

"Truth or dare?" Pete asked.

"It seems everybody loves to do dares, so I choose it too," Oliver replied.

"I dare you to kiss Andrew for thirty seconds," Pete challenged Oliver.

"Why? Didn't you hear Hassan?" Andrew did not like the dare given Oliver.

"Just do it," Hassan said.

Oliver moved and sat on Andrew's lap; he held his head in his hands and kissed him for thirty seconds. Oliver moved after the kiss, and Hassan hit Andrew with a pillow.

"We didn't want to see that," Hassan said.

"I am sorry," Andrew blushed.

Oliver spun the bottle and pointed to Kyle.

"Truth or dare?" Oliver asked Kyle to choose.

"Dare is more fun, so dare," Kyle responded.

"I dare you to dance like a ballet," Oliver said.

"Okay, I can do that," Kyle happily agreed.

Kyle danced like a ballet, and all laughed. They kept on playing, and all enjoyed it. For years, Oliver hadn't laughed like that. Andrew looked at Oliver and wondered, "Who's that man?" They had kissed for a few seconds and the erection that he had during the kiss was his thought.

Everyone went to bed after the game. Oliver stayed in Hassan's room and they stayed up late to talk. Andrew, on the other hand, couldn't sleep. All he could think of was the kiss. He thought he ought to go to Hassan's room. He stood in front of his door for a while. Then he finally knocked.

"Come in," Hassan said and Andrew entered the room.

"Can we talk a little?" Andrew asked Hassan.

"Sure," Hassan said.

"What do you want to talk about?" Oliver asked Andrew.

"Isn't it obvious?" Hassan implied something to Oliver.

"I have no idea what you are talking about?" Oliver was confused.

"It's about the kiss, right?" Hassan asked Andrew.

"Right." Andrew nodded.

"It's just a kiss." Oliver thought as if it was normal.

"Is it?" Hassan asked.

"All I can think about right now is kissing you," Andrew confessed to Oliver.

"I don't think it's a good idea." Oliver was not interested.

"Shut up and let me kiss you." Andrew jumped him.

Andrew kissed Oliver. Oliver looked at Andrew and he tried to say something, but he lost his words. Oliver was panting.

"Is he alright?" Andrew asked.

"He just panicked. He hasn't kissed anyone for years." Hassan tried to calm both of them down.

Andrew looked at Oliver and told him, "That kiss was perfect."

"You think?" Oliver asked.

"I had two perfect kisses in my entire life. You are both good kissers," Andrew said.

"You've kissed Hassan?" Oliver was shocked.

"It was nothing, just a kiss." Hassan clarified.

"Oliver, why are you sad?" Andrew asked.

"It's not the time." Hassan tried to divert the conversation.

"It's okay. I am going through a rough patch. I hope it will clear up soon," Oliver replied.

"I hope so," Andrew said.

"So, will you sleep here with us?" Hassan asked Andrew.

"I would love to. You have a comfy big bed," Andrew said.

"Yours is too," Hassan directed towards Andrew's bed.

"Don't be greedy. Let me stay here for the night." Andrew stopped Hassan from saying anything further.

"Leave him be," Oliver said while smiling.

"Okay, but if I see your fingers, and I mean any finger touching me. I will cut it down. Deal?" Hassan set the rules.

"I am scared. I will sleep next to Oliver if he doesn't mind," Andrew tried to sneak in with Oliver.

"Not at all." Oliver showed no reservations.

Andrew and Oliver stayed in the same bed with Hassan that night. He moved Oliver between his arms before he fell asleep. Oliver woke up in Andrew's arms.

"You slept like a baby," Andrew said to Oliver.

"Morning. I hadn't had a good sleep like this in years." Oliver was relieved after having a good night sleep.

"Morning. Really! Why?" Andrew said.

"I would say I felt safe," Oliver said.

"Between my arms," Andrew specified.

"In this house," Oliver clarified.

"Who are you?" Andrew asked.

"I am Oliver, Hassan's friend."

"You are mysterious, aren't you?"

"Believe me, you don't want to know me."

"Why is that?"

"I am too much to handle."

"Try me!"

Oliver got up. He put on a pajama. While doing that, Andrew saw a burn on his leg. He got up and asked Oliver, "How did your leg get burnt?"

"This is my secret."

"It must have been horrible for you."

"You have no idea." Oliver cried.

Andrew went over and hugged Oliver. "The symbols of the life you have lived to fight to be here are the scars you carried with yourself. But I can see that you are a strong person, I don't know what your story is, but you are strong," Andrew said.

"You think I am strong?" Oliver asked.

"Takeaway to zoom out and get the big picture of your life. You will find yourself very damn strong." Andrew tried to excite Oliver.

Andrew looked at Oliver's face, and they kissed.

"We have to go down before Hassan becomes a morning monster," Andrew told Oliver about Hassan's morning routine.

"He is so friendly, don't say things like this about him." Oliver did not agree with Andrew.

"Believe me, he is evil," Andrew said.

Oliver smiled and Andrew said, "There goes a big fat smile." Andrew kissed Oliver again.

After breakfast, Hassan took Oliver aside to talk to him.

"How are you feeling today?" Hassan asked.

"I am feeling better. I haven't slept for years like this." Oliver replied with relief.

"You slept like a baby between Andrew's arms."

"He is nice."

"You can stay as long as you want to."

"Thank you."

Hassan and Oliver finished their conversation, and Andrew walked to Hassan to talk to him.

"Can we talk?" Andrew asked Hassan.

"Sure," Hassan replied.

"Who is he?"

"Do you like him?"

"Yes, but I don't know anything about him."

"What did you do to make him smile?"

"We talked and kissed."

"That's all?"

"Yeah. You didn't answer my question."

"You are asking the wrong person. If you want to know who he is, ask him."

"Why did you ask us not to mention him to anyone?"

"I am protecting him."

"From whom?"

"Why don't you ask him yourself?"

"Why are you so weird?"

"Andrew, Oliver is a person who is more than you can handle. He has been through a terrible tragedy."

"Why do you think he is more than I can handle?"

"Do you want to know why?"

"Yes, please."

"Fine, I will tell you tonight, but never tell anyone."

"You know me."

"I know I can count on you."

"So why are you being a pain in my ass?"

"You will understand soon. Now, I have to get to work."

With many questions, Hassan left Andrew, but he'd have the answers by the night as he was curious about Oliver.

Hassan took Oliver and dropped him near his office. Ethan saw Hassan and caught up to him to ask, "Who's Oliver? I know some details about him, but I'm not too sure."

Hassan looked at Ethan and said, "I know. He needs my help. I'm giving him his space."

"I trust you. I just want you to be careful." Ethan tried to assure Hassan of his support.

"Thank you, I will." Hassan appreciated Ethan's candor.

Hassan and Ethan arrived at work. He asked Hassan if he was ready for the meeting.

"I am ready," Hassan replied.

Hassan was concerned about the presentation before the meeting. He was feeling excessive pressure. He went to the conference room and watched Klaus and Nina walking with the department head, Michael.

Chapter 18: A Serendipitous Journey

"Some people say coincidences are meaningful. That it's the universe trying to tell you something" - Lourelin Paige

Hassan was watching Klaus from afar, who was coming toward him in a blue suit, smiling and gliding. Michael stopped and brought Klaus and Nina to Hassan. He was there.

"We know each other. Right, Hassan?" Nina asked.

"Right," Hassan replied.

"Why haven't you told me?" Michael asked.

"I wasn't aware of our contract with Klaus and Nina's company." Hassan justified.

"Are you close?" asked Michael.

"Yes, I am Hassan's boyfriend," Klaus replied.

"Oh, that is wonderful. Now let start our meeting." So, Michael proceeded to the meeting.

"We will catch up with you," Klaus said.

"Right behind you," Nina followed.

"Are you okay?" Klaus asked Hassan.

"This is too much pressure for me," Hassan exclaimed.

"Is it?" Klaus smiled at Hassan.

"Stop looking at me like that. You are not helping." Nevertheless, Hassan was still anxious about the meeting.

"We have to start, shall we?" Nina began the meeting.

Klaus and Nina entered the meeting room, and Hassan followed behind them. Hassan began his presentation, and everyone was listening to him attentively. He was a little worried, but soon, he got over it. He demonstrated confidence and presented information with the most effective approach, and his data was well-researched; he presented it in such a way that everyone understood him. After Hassan finished his presentation, many questions were asked, and he answered all of them. Klaus and Nina smiled after everyone was finished with questions.

"That was very impressive," Nina commended his presentation skills.

"He is Hassan, he is one of the best on our team," Michael backed Hassan.

"We can see that," Klaus seconded.

They started to talk about the proposal, which ended with an agreement. Nina looked and smiled at Hassan.

"So, that is Hassan. Finally, someone good enough to beat Klaus's habits," Nina said to herself.

Nina wanted to have a private conversation with Hassan after the meeting.

"Can we talk?" Nina asked Hassan.

"Sure." Hassan responded.

"Who are you?" Nina asked out of curiosity.

"I am Hassan. We met yesterday," Hassan replied with a smile.

"I know your name, but I don't know how you managed to be Klaus's boyfriend. You achieved the impossible," Nina stressed Hassan impressing Klaus to be his boyfriend.

"Klaus is a charming guy, and I really like him," Hassan replied calmly with a calm candor.

"Have you met his kids?"

"I've met them all."

"I haven't seen him smile like this in years. I want to say thank you."

"Can I get my boyfriend back?" Klaus intervened in the conversation between Hassan and Nina.

"He is all yours," Nina replied with a soft chuckle.

Nina left Hassan with Klaus.

"What did she say?" Klaus asked.

"She wants to know me," Hassan replied.

"You were great there," Klaus said.

"You think?" Hassan asked.

"Can we have lunch together?" Klaus asked Hassan if he would go out for lunch with him.

"I have work to do," Hassan excused.

Klaus turned to Michael and directly asked for his permission to take Hassan out for lunch, "Michael, is it okay to take my boyfriend to have lunch together?"

"He is all yours. Excellent work, Hassan," Michael approved while lauding Hassan's work.

"Thank you," Hassan acknowledged.

"So, now that you are free," Klaus said, giving Hassan a teasing look that he can't refuse him now.

Hassan, Klaus and Nina walked toward the elevator. Klaus whispered in Hassan's ear, "Why don't you kiss me?"

"We are in the elevator," Hassan said.

"Come on, kiss him already," Nina insisted.

Hassan and Klaus kissed, and Nina smiled. Then, Klaus brought Hassan and Nina together to a restaurant for lunch. Klaus held Hassan's hand and went in together when they arrived at the restaurant. Nina looked at them with a big smile on her face.

****

Andrew asked Hassan to speak privately at night so, he went to his room.

"Here we are, you can tell me," Andrew kept a low tone.

"Promise me whatever I tell you will stay between us," Hassan put a condition for Andrew to keep the secret to himself.

"I promise. Now tell me," Andrew said.

Hassan started narrating Oliver's story, and Andrew listened to it quietly, which was unusual of him. What Hassan said wasn't possible for Andrew. While listening to Hassan, Andrew felt intense pain. But he stopped his tears from falling, and he was about to cry. Andrew put his hands on his mouth after Hassan told Andrew and began to cry. Hassan hugged Andrew, "It's okay. Bad things happen to good people all the time," Hassan said.

Andrew was unable to control his tears. He felt a mixed storm in his body, which he could no longer contain. He would stop, wipe his tears, only to break down again; this went on for 15 minutes. "How can a person go through all of that and still stand?" Andrew said.

"Carrying that pain all that time doesn't make you weak, but it makes you unstoppable if you can manage to stand up again." Hassan said to Andrew.

"What are you going to do?" Andrew asked while slightly panting.

"I have to stand by his side till the end," Hassan answered Andrew.

"That's why you are my favorite person in the whole world," Andrew rushed to hug Hassan.

"Come here," Hassan pulled Andrew towards himself.

Hassan hugged Andrew and kissed him on the forehead. "You didn't tell me about your new boyfriend," Hassan said.

"He is John's therapist," Andrew told Hassan about the guy he was dating.

"Who?" Hassan forgot about the guy Andrew was talking about.

"John's therapist. Do you know him?" Andrew asked.

"I don't know him, but since John started to see him, nothing good happened. He is a terrible shrink," Hassan showed his disapproval.

"Why?" Andrew asked.

"He is the one who told John to walk away from Carlos when they had a problem." Finally, Hassan revealed the truth behind John's breakup.

"How did you know?" Andrew was intrigued.

"John told me, then you saw what I did. So, I'd say dump him now," Hassan told Andrew.

"Why do you want me to dump him?"

"Do you want him to ask you after every time you had sex, 'How does that make you feel?'"

"You are right. That would be weird. It will be turned into a therapy session."

Andrew and Hassan were laughing. Andrew went in between Hassan's arms. "Are you feeling good?" Andrew asked.

"I am happy," Hassan replied.

"Are you falling for him?"

"I am, but this time, it's different.

"What do you mean?"

"I've loved John for six years, but my feelings for Klaus are very strong. I feel I belong with him, and when I am with him, my body yearns for him in a way I cannot resist. I am thinking about him all the time."

"I am happy for you."

"I am terrified."

"Of being hurt?" asked Andrew.

"I don't want to get hurt."

"Klaus is different. I am sure he is the one."

"What if he isn't the one?"

"If he is not the one, you end up with me."

"Your chance is more significant than mine to have a boyfriend."

"Who wouldn't be with you? Believe me when I tell you Klaus is the one."

"How do you know?"

"He cannot get his eyes off you. The way he looks at you. I can tell he is different."

"You are my best friend."

"I know."

"I want you to promise me something."

"Sure."

"Whatever happens, I want you to keep my family together."

"Are you going to die?" Andrew asked sarcastically to ease the tension.

"Just promise me."

"I promise. Now, can you tell me what is going on?"

"I know it sounds like a drama, but every time I feel happy, something comes up, and catastrophe happens and everything is washed away."

"I want you to promise me something?" Andrew asked Hassan for something while in tears.

"Sure," Hassan nodded.

"Never ever think about killing yourself again."

"I know I did it once, and I won't do it again," said Hassan.

"Promise?"

"Promise."

Someone knocked on the door, and it was Ethan. Hassan told him to come in.

"Hassan, you need to come down," Ethan asked Hassan in a rather worrying way.

"Is everything okay?" Hassan did not have a good feeling about this.

"Please, come right now." Ethan implied urgency.

Hassan stood up and went downstairs with Ethan. Hassan felt that something was going on that profoundly hurt somebody. Hassan looked over everyone, and he saw Tristan wearing the jacket of his father and crying.

"He didn't say anything, but just that he wants to talk to you," Ethan said.

Hassan ran to Tristan and hugged him. "It's okay now. You can let it out," Hassan said.

"I don't know what to do," Tristan cried and hugged Hassan.

"We will figure it out, but how are you holding now?" Hassan asked.

"I lost everything," Tristan kept on crying.

"That is not true. You have me. You have us. He was very proud of you and the man you've become. You need to let me take care of you. Ethan, can you please take him to my room. I will be there, I have to do something before that," Hassan said.

While Hassan had gone into the kitchen to prepare a soup, Ethan took Tristan to Hassan's room.

"What happened? Why is he crying? And what are you doing?" Cara asked everyone there.

"His father passed away, and I am preparing soup," Hassan disclosed the sad news to Cara.

"Soup!" Kyle said.

"I am preparing Tristan's father's recipe," Hassan told the others.

"I can help," Pete stepped forward.

"Thank you. One more thing, he will stay here for some time. Is that okay with you guys?" Hassan asked if anybody had objections to Tristan staying over.

"As long as he wants to," Andrew showed a welcoming attitude.

Andrew looked up to Hassan and saw that his face maintained a troubling expression. He could not understand why Hassan had that look on his face. Hassan finished the soup, poured it into a bowl and brought it upstairs to Tristan.

"Hey, you need to eat something," Hassan said, putting the tray on the table.

"I can't," Tristan was crying and was unable to eat due to grief.

"You can. It's your father's recipe," Hassan said.

"Is it?" Tristan paused and then asked Hassan if he was telling the truth.

"I made it myself with Pete's help," Hassan confirmed.

Tristan took the tray and had that soup from Hassan's hand. Hassan hugged Tristan and put the tray away. "Do you know what my father would say about your soup?" Tristan asked.

"He would say, 'Child put some salt in it. You haven't used any,'" Hassan replied.

"That is exactly what he would say. Thank you," Tristan smiled.

"You don't need to thank me, we are friends," Hassan tried to get over the formality.

"Can I stay with you?" Tristan asked.

"As long as you want," Hassan replied.

"I meant moving to live here."

"You will stay in my room for a while until I prepare the room upstairs for you."

"My father wanted me to be here, between all of you. He was always fond of you."

"He was a great man. He raised you to be a responsible man."

"Then, why am I crying like a girl?"

"Did you meet your mother today?"

"She was at the hospital when my father passed away. She said that even though I look like a man, on the inside, I am a girl like the way I am crying."

"And you listened to her?"

"Maybe she is right."

"She is wrong. You are a man inside out, and nothing wrong with crying. It is the way normal people react when they lose someone. The only thing wrong is that you listened to her."

"She is wrong!" Tristan exclaimed.

"Yeah, she is. You know, your father told me once that you became the son he always dreamed about."

"He did?"

"I wish you can see yourself through your father's eyes right now. You were the most important person in his life."

"Thank you."

Andrew knocked on the door, "Come in," Hassan said.

"I want to check on you. How are you feeling?" Andrew inquired about Tristan's condition.

"I feel better," Tristan replied.

Andrew laid down on the bed next to Hassan, "I want to tell you something," Andrew said.

"Are you talking to me?" Tristan asked Andrew if he was talking to Tristan.

"Yes," Andrew said.

"Spill it!" Hassan maintained a hard tone.

"We might eat your soup. All of it," Andrew said subtly and slowly. Andrew then laughed with Hassan and Tristan.

****

Hassan was with Charley a few weeks later at the shopping center. Charley was extremely worried as she had left the baby with María.

"Stop worrying. She is with María; you basically have been raised by María," Hassan tried to assure Charley.

"You are right," Charley agreed with Hassan.

Someone came by and said, "Hi" to Hassan. Hassan looked at him and said, "I know you from somewhere, but I cannot remember."

"We met a few times. I am Anton. I am working with Nicholaus and Nina," Anton introduced himself to Charley and Hassan.

"Oh yeah, but we never talked before," Hassan remembered.

"Who is the lovely lady with you?" Anton was charmed by Charley.

"She is Charley, one of my friends," Hassan introduced Charley to Anton.

"I am Anton. Nice to meet you," Anton brought his hand forward to Charley and introduced himself formally.

"Nice to meet you," Charley replied.

Anton couldn't take his eyes off Charley. He found himself attracted to her.

"We were about to grab a bite. Would you like to join us?" Hassan asked Anton.

"If that is okay with you too." Anton was happy with it.

"We would like that," Hassan told Anton.

Hassan, Charley and Anton sat at a table and tried to decide what they wanted to eat. Anton insisted that he receive Charley's order. Hassan and Charley had a little chat while Anton was waiting for their orders.

"I am your friend?" Charley was not interested in Anton joining her and Hassan.

"Didn't you see the way he looked at you?" Hassan tried to convince her.

"He is older than me, at least fifteen years," Charley argued.

"So? I am dating your father, and he is older than me."

"Dating is not my game anymore. Now I am a mom," Charley said.

"You are pretty and young. Anyone will be lucky to be with you. Plus, he is really handsome," Hassan kept on convincing Charley.

"He is charming, but I don't think he will be into me after knowing I am a single young mom." Charley was sure about what she said.

"How do you know when you haven't even tried?" Hassan asked.

"I have never been with someone older than me," Charley answered.

"He will be your first," said Hassan.

Anton came with their order and sat down to eat. Hassan was watching the way Anton was looking at Charley.

"Anton, are you single?" Hassan asked Anton if he was committed.

"I am a widower. My wife died five years ago."

"And you haven't been with anyone since?" Charley asked Anton.

"I couldn't find anyone," Anton replied.

"What do you mean?" Hassan asked.

"It is like I am searching for this perfect woman, who does not exist," Anton replied.

"Do you have kids?"

"Unfortunately, no."

"You know, Charley is a single mom. She has Peyton, a lovely girl."

"Oh, I didn't realize you were married," Anton was surprised about Charley being married.

"I am a single mom. Thanks for bringing this up." Charley replied, annoyed at his assumptions.

"I think you are strong and brave enough to have a baby. I am sure you will find a good father for her," Anton gave well wishes to Charley.

"That's what I told her, and she is always talking about how no one wants to be with a single mom," Hassan talked about Charley.

"That is not true. You are a pretty rare flower; anyone will be lucky to have you," Anton emphasized Charley's beauty.

"You think so?" Charley blushed.

"I would date someone like you, especially after knowing you are a single mom. But I don't think you are into someone my age." Anton kept on impressing Charley.

"See, Anton agreed with me. And Anton she thinks that age is just a number," Hassan told Anton.

"Stop it!" Charley said to Hassan as she couldn't stop blushing.

"Would you like to go out with me on a date?" Anton asked Charley out.

Charley looked at Hassan, "Why are you looking at me? He is the one who asked you, and I am not his type," Hassan teased her.

"Take your time to think," Anton did not want to force.

"She is a little shy, but believe me, she would," Hassan talked on behalf of Charley.

"I want to hear from her," Anton preferred Charley agreeing by herself.

"If you don't date him, I would," Hassan whispered in Charley's ear.

"I will tell my dad about you," Charley whispered back.

"She said yes, but she is shy," Hassan replied.

"I would like to go out with you," Charley said yes to Anton's proposal.

Anton took Charley's hand and kissed her hand softly, "I promise you I will treat you the way you deserve, my lady," Anton said.

"Oh, he is a gentleman," Hassan complimented Anton's demeanor.

They had a little chat and finished their meal. After that, they decided to continue their visit to the shopping center for a while. Anton had locked his eyes on Charley, he was so into her, and he did not realize that yet. He smiled, thinking, "She is really a wonderful lady. She is everything I want and has a daughter. She is perfect." They were going home after they were done shopping. Anton and Charley agreed on Saturday night for a date; Anton took Charley's phone number. Then Hassan and Charley left on an Uber.

"Do you think he is into me?" Charley asked Hassan.

"He wants to be a father for your daughter," Hassan guaranteed Charley about Anton.

"How do you know?" Charley asked.

"I am like your father," Hassan replied.

"Do you think that we will have an intimate moment on the first date?"

"Maybe fourth or fifth date."

"Really!"

"The man kissed your hand, and he would treat you like a lady."

"I am not quite sure; what do you mean?"

"He is a gentleman. He wants to know you before anything happens between you both. I take it you have never been with an old-fashioned gentleman. I hope you are into him because a man like that is not easy to find."

"You are saying that I have to date him?"

"You have dated a few men your age, right?"

"Right."

"Did you like dating someone your age?"

"Not really."

"Then, you will enjoy dating an old-fashioned gentleman," Hassan assured her.

"I am afraid."

"That is normal."

"I am happy to have you in our life."

"I might become your stepdad."

"You would be a wonderful stepdad." Charley didn't take notice of his teasing.

Hassan hugged Charley like his own child. Charley took a deep breath and smiled. "I am afraid that my dad won't be happy about me dating Anton," Charley said.

"You won't tell him until we are sure that you are into Anton. For the time, leave your dad to me. I can handle him."

Hassan and Charley went home. Hassan ran into Adam the moment he opened the door. Hassan hugged Adam, sat on his knees and kissed him. Adrien walked to Hassan and said, "Hey, you are home," and hugged Hassan.

"You missed me? I was here last Sunday," Hassan chuckled and replied to Adrien.

"He wants to tell you about his day at school," Edouard told Hassan.

"So, are you?" Hassan asked.

"I did very well today on the exam," Edouard said with sheer happiness.

"You did your part of our deal. Now, it's my turn," Hassan said.

"You are spoiling them," Klaus said to Hassan.

"They are doing great, and they deserve a reward," Hassan replied.

"Do I deserve a reward?" Klaus asked.

"Hmm, sure," Hassan nodded.

"So, tell me, how was your day with Charley?" Klaus asked about their lunch.

"It was great. We had so much fun, and finally, she let me pick a dress for her," Hassan said with joy.

"Can we see it?" Klaus asked.

"Dad!" Charley smirked.

"I want you to wear the dress and show it to us," Klaus asked Charley.

"We only see you in jeans. I don't remember I have ever seen in a dress," Edouard insisted Charley wear the dress too.

"I will help her get dressed," María pledged to help.

Charley came to her room to put on the dress. Adrien began talking to Hassan about his school day, and Hassan listened to the kids with a big smile on his face. Klaus held Adam between his arms and took a break, with Adrien watching Hassan. On his father's face, Edouard saw a smile. He smiled and hugged his father, "I love you, and thank you for bringing him to our life," Edouard said.

"Can I tell you a secret?" Klaus asked.

"Sure," Edouard replied.

"Your mother likes him," Klaus said.

"I know," Edouard said as if he knew.

Klaus looked at Edouard and smiled, "I missed you, son. I would never hurt you again," Klaus said.

Edouard looked at his father and smiled, "I want a future with him, dad," Edouard said.

Klaus took a deep breath. Charley walked in her splendid dress with María. Klaus looked at her with his jaw dropped. He couldn't believe what he was seeing.

"You look amazing," Klaus was stunned.

"Dad!" Charley was happy with the compliments.

"You look lovely in that dress. You are the spitting image of your mother in that dress," Klaus was happy, and he was holding back tears.

"Wow, you are a real lady," Adrien was amazed.

"Yeah, that's my girl. She is breathtaking," Hassan praised Charley's beauty.

"I already saw myself in the dress," Charley said.

"And you are so beautiful. Would you like to dance with me?" Hassan put his hand forward to Charley for a dance.

"If you insist," Charley replied.

Edouard was playing music, and Hassan took Charley's hand and began dancing. Klaus watched Hassan with a smile on his face. María watched Charley with unbridled happiness and told herself, "God answered my prayers."

Edouard walked to Charley and asked Hassan if his sister could dance with him. Smiling, Hassan said, "Sure." María was asked by Hassan to dance with him. Bella began to dance to Adrien. Klaus had not felt this joy and warmth for such a long time. He shut his eyes and breathed deeply. "I really love him," Klaus told himself. María took Adam from Klaus after a couple of minutes so he could dance with Hassan.

Hassan looked at Klaus, "I've never felt like that before," and it was all he thought. Klaus moved his face slowly, breathing on it, giving a deep passionate kiss to Hassan. Hassan felt his heart stop with the whole world at this moment. He was deeply in love with Klaus. Klaus gave him all he wanted. Hassan shut his eyes, put his head on Klaus' chest and took a long sigh. Everybody was happy in that moment. Charley, Adrien, and Edouard knew they had two dads now. Hassan, the foreigner who came into their lives, changed it all after their mother died.

The strange feeling that Lizzie was with Charley, Adrien, Edouard and Klaus. María looked at the piano, where Lizzie usually sat. When she saw Lizzie going to Hassan, she smiled. From the back, Lizzie took Hassan and placed his hands on Klaus' back. "Thank you for bringing back my family," Lizzie said.

Hassan felt he had been in heaven, as his family was around him and the entire world was spinning. Klaus sighed as he felt his heart taken captive by love. His soul had found love with a person who was strong enough to collect and heal the pieces of his broken heart and return a balance to the family. The picture now differed. Klaus watched Hassan in the big picture. Strangely, chilling, Hassan felt a sigh, holding Klaus narrowly.

****

A few weeks later, Hassan went to work at the restaurant. But he couldn't find Ruby. He was concerned about her, so he asked the other restaurant worker, "If Ruby comes, just ask her to come to my office."

A while later, Ruby came, and she was told to see Hassan in his office. Ruby was concerned that Hassan would be mad about her and fire her. So, she entered the office and said, "You asked for me?"

"Yes, please have a seat," Hassan invited her in.

"I am sorry for being late. Please don't fire me," Ruby requested.

"I'm not going to fire you," Hassan cleared Ruby's doubts with a soft smile.

"Why am I here?" Ruby asked.

"I wanted to check on you if you have any problems or you need something?"

"You have done enough for me."

"I didn't do anything. If you want to reduce your work responsibility or work time, it's okay. You can ask."

"I just need to figure out things with my mom about our schedule. After that, I promise you I won't be late."

"How much do you love your kid?"

"More than anything." Ruby was in tears.

"Then let me help you. There is a daycare center close to the restaurant. I will take you there to register your baby," Hassan advised Ruby

"I cannot afford it."

"I know you want the best for your kid. So just say yes."

"Thank you."

"One more thing. We didn't have a baby shower for you, so we saved the money, and I will send it today to your account."

"You are doing a lot of things for me. Thank you."

"You did choose to fight for your kid, and I will make sure that you fight till the end."

"I don't know how to thank you.," Ruby sighed.

"No need to thank me. Let us go to the daycare center," Hassan comforted Ruby.

Hassan went with Ruby to register her kid at the daycare center. Ruby was glad that she had Hassan on her side. She said to herself, "He will be an amazing father." When they got back to the restaurant, Hassan saw Carlos waiting for him.

"Hey, Carlos. What a surprise?" Hassan said to Carlos.

"Hey, how are you doing?" Carlos greeted back.

"I am doing great, thanks for asking. How are you doing?"

"I am doing okay, thanks. Do you have a minute?"

"Sure, let's go to my office."

Hassan took Carlos to his office to talk. "How can I help you?" asked Hassan.

"Is the job offer still on?"

"Of course."

"When can I start?"

"Can you tell me why have you changed your mind?"

"I quit."

"And the reason?"

"My salary was not enough and my schedule has been changed."

"Okay, do you have a resume?"

"I didn't have time to print it."

"Can you send it to me?"

"Sure."

Hassan gave his email to Carlos so that he could send his resume to Hassan. Hassan printed it when he received it.

"Can I get your ID, social insurance number and your bank account information?"

"I have it with me."

"Great, now I'll walk you through your salary, benefits, work hours and tips. But first, I need to know your schedule."

"You are paying benefits?"

"Why do you look surprised?"

"It's just I haven't received any benefits before. Anyway, I won't be able to work today. I have school."

"No worries."

Hassan wrote per hour rate for Carlos on the paper and showed it to Carlos, and then he discussed all of the details. "Now, you will work four days a week. Is that okay for you?" Hassan said.

"Why not five days a week?"

"You need a day off to plan your week, and you need to be with your boyfriend."

"I don't know how to thank you."

"I know a way."

"Okay!"

"I want you to be happy."

"Thank you. Now I can see why John is crazy about you."

"Was. Now, he is crazy about you. You gave him another chance to be happy."

"Whatever happens, please don't take his side without listening to me."

"I am listening."

"He slept with many people in the past, and that terrified me. What if he slept with more or left me for another one?"

"He is not that kind."

"Then, why did he do it in the past?"

"He didn't tell you?"

"Yes, he did."

"He is not a bad guy. He is just a man with feelings, and he would never do the same to you."

"How do you know?"

"Because he has learned his lesson and won't make the same mistake again. Oh, my God. You are in love with him."

"I don't know what to do?" Carlos' tears started dripping down his eyes.

"Don't be your own enemy."

"What do you mean?"

"You have all the tools to be happy. Use them to be happy. I want John to be happy, and you are perfect for him. So, stop being a baby and find a way to be happy with him."

"Got it," Carlos smiled.

"Now go, and Ruby will show you around."

Carlos went around with Ruby. John called Hassan, and the first thing he asked was about Carlos.

"Hey. Did Carlos come to you?" John asked Hassan about Carlos' visit.

"Yes, he is here. Can I tell you something?" Hassan said to John.

"Sure," John replied.

"Don't ever hurt him, or I will kill you."

"Can you tell me what happened?"

"He is in love with you, dummy."

"Really!" John was happy.

"He is your chance to have everything you ever dreamed about. Don't make the same mistake again."

"I will ask him to move in with me. Is that okay?" John took a long sigh.

"Why are you asking me?" Hassan asked John.

"It is your house."

"Do you think I will stand in the way of your happiness? Plus, he is one of us now."

"Thank you."

"Just be happy. Now I have to go. See you at home."

"See you at home."

The call ended. John placed his phone on his heart and breathed deeply, "He is in love with me," John said.

After work, Hassan returned home and found Andrew eating ice cream. "What happened to you?" Hassan said.

"Dating a shrink sucks. I am like people who have been thrown into the emotional wringer," Andrew shouted.

"I thought you were done with him?" Hassan was confused.

"I lied, okay. Every one of you has a partner except me. I am freaking alone. Don't you see me crying on the inside?" Andrew was in tears while eating ice cream.

"What do you want me to do? I set you up with some dates, but you weren't interested," Hassan replied.

"You know what you have to do," Andrew told Hassan.

"We've talked about this many times," Hassan took a long sigh.

"And yet you don't get it., Andrew replied instantly.

"You know it's wrong." Hassan tried to correct Andrew.

"You are wrong," Andrew quickly blamed Hassan.

Hassan took his phone and called Oliver.

"Hey, Oliver. How are you?" Hassan greeted.

"I am fine, thank you. How are you?" Oliver replied.

"I am okay. Can I ask you something?" Hassan asked.

"Sure," Oliver answered.

"Are you interested in dating Andrew?" Hassan talked on behalf of Andrew.

"He is a nice guy, but I don't know how he will react if he knows my story," Oliver voiced his concern.

"I told him, and he doesn't mind. He has asked me a thousand times about you," Hassan tried to convince Oliver.

Andrew took Hassan's phone, "What the hell?" Hassan said.

"Just tell me, yes or no." Andrew took matters into his hands and asked Oliver bluntly.

"I think I have no choice," Oliver replied.

"Is that a, yes?" Andrew tried to confirm.

"Yes, I would like to date you," Oliver instantly said.

"I will take your number from Hassan," Andrew said to Oliver.

Andrew gave the phone back to Hassan. "He said yes," Andrew said.

"Okay, Oliver. Talk to you later," Hassan said to Oliver.

"Bye," said Oliver, and the call was cut.

"Are you happy now?" Hassan said to Andrew.

"Shut up and be happy for me. I promise you I will be very careful," Andrew gave a shut-up call to Hassan.

"I know you will," Hassan trusted Andrew's words.

"By the way, John accidentally mentioned Oliver to his shrink," Andrew told Hassan about John's actions.

"I don't like that shrink. Anyway, Oliver is not his real name. He changed it recently."

"What is his real name?"

"Never ask this question, understood?"

"You are evil."

"I know."

"John will ask Carlos to move in with him."

"I know. It's about time that John follows the way to his happiness."

"What about Klaus?"

"What about him?"

"Do you love him?"

"I am deeply in love with him."

"Does he feel the same?"

"What do you think?"

"I see him crazy about you. You bring out the best of him, and he is always melting in your smile. It's about time to be happy."

"I am happy, but…."

"What?"

"I have that feeling that someday not far, something will happen."

"Don't think like that; he is in love with you."

"I know. What do you think about inviting him and his kids for dinner on Saturday?"

"That is a good idea. Please don't invite the shrink."

"I am not insane to invite him, but I want to invite Oliver. Is that okay?"

"I don't mind. I will call Klaus and let him know."

"Before that, where is Oliver's number?"

"I will send it to you."

"Okay."

Hassan sent Oliver's phone number to Andrew, and then he called Klaus.

"Hey, you. How is it going?" Hassan asked Klaus.

"Everything is okay. What about you?" Klaus replied to Hassan.

"Everything just fine. I was wondering if you, the kids and María could have dinner at my house on Saturday with the whole gang?" Hassan invited Klaus and his family to dinner at his place.

"With the whole gang?" Klaus asked while laughing.

"You know, the gang at my house," Hassan replied with a laugh.

"I know what you mean, and we would love to join you and the gang for dinner on Saturday. But now you need to save me," Klaus agreed but asked for a favor in return.

"What is happening?" Hassan asked.

"My parents are here in Toronto, and they want to meet you," Klaus informed Hassan about his parents' arrival.

"Your parents?!"

"Yes, and they are here to meet you."

"Why? I mean, why?"

"They want to meet my boyfriend, that's all."

"Okay, I will go to your house. I have to check on Peyton anyway."

"What about Peyton? Is everything alright?"

"Relax, nothing happened. Just Charley has a date today, and I offered to take care of Peyton."

"Charley has a date?!"

"Did you think she will stay single forever?"

"I didn't see that coming. Anyway, who is the lucky guy?"

"I have to go. See you at home."

"But…" Klaus was about to say something, but Hassan cut the call.

Hassan didn't tell Klaus who Charley was going on a date with. But he was about to tell Klaus soon. Hassan took a shower and then rushed to Klaus' house. María opened the door.

"Thank God you are here," María was relieved to see Hassan.

"Sorry for being late," Hassan apologized for the delay.

"Hassan, am I right?" Valerie (Klaus' mother) asked.

"Yes, I am Hassan. Nice to meet you." Hassan replied.

"Likewise." Valerie shared mutual feelings.

"Klaus didn't mention that you are so pretty," Hassan complimented Klaus' mother.

"Thanks, my dear." Valerie appreciated the compliment.

"So, you are the one who's Nicholaus dating?" Klaus' father, Denis, asked.

"Hi, sir. Yes, I am Hassan." Hassan moved forward to meet Klaus' father.

"Please call me Denis." Denis asked Hassan to spare the formality.

"Nice to meet you," Hassan replied.

"We heard so much about you," Denis said to Hassan.

"We are curious about you," Valerie said.

"I hope you heard good things," Hassan replied.

Adam walked to Hassan. Hassan picked him up and played together with him. Adam was rattling.

"The kids love you. No wonder why," Valerie acknowledged the kids' love for Hassan.

"He treats them well," Denis seconded Valerie.

"They are a part of my family, and I would never hurt them," Hassan replied humbly.

"Charley told us you picked the name for her daughter," Valerie asked.

"I suggested the name, and she liked it," Hassan replied.

"Grandpa and grandma here. I am so happy to see you." Edouard came rushing towards his grandparents.

"I missed you so much." Soon, Adrien came in and hugged his grandmother.

"How is school?" Denis asked Edouard.

"I am doing great at school. Everyone loved my science project," Edouard told his grandparents about his day at school.

"A science projects!" Valerie exclaimed in surprise.

"Yes, Hassan helped me with it," Edouard replied.

"I see. Welcome back, little nerd," Valerie was happy.

"Where is Charley?" Denis asked about Charley.

"She has a date," Adrien answered.

"A date!?" Valerie seemed to be surprised.

"Yes, that's true," Edouard said.

"Hassan, do you know about Charley's date?" Valerie asked Hassan.

"Yes, I know everything about it," Hassan replied.

"But you never told me who she is dating. Hi, mom and dad," Klaus came in and greeted his parents.

"Hey Klaus," Valerie replied to Klaus.

"Hey, son," Denis also greeted his son.

Klaus kissed his mom on the cheek and forehead and hugged his father.

"Can you tell me who she is dating?" Klaus asked Hassan about Charley's date.

"Promise me you will stay calm," Hassan asked Klaus.

"I will try," Klaus gave an uncertain response.

"Edouard, can you take Adam and Adrien with you?" Hassan asked Edouard to take Adam and Adrien to the room.

Edouard seized Adam and took Adrien out of Hassan's hands, and went upstairs.

"Who is she dating?" Klaus asked Hassan.

"She is dating Anton, the one who works with you," Hassan finally disclosed.

"Who? Why? He is older than her," Klaus did not take it positively.

"She likes him, and she is a grown woman. She can choose who she is dating," Hassan tried to justify.

"No, I will not allow this to continue," Klaus replied.

"Why?" Hassan asked.

"You asked me to get more involved with my kids' life, and she is my kid."

"I said to get more involved, not take control of their life."

"Why are you doing this? They are not your kids," Klaus yelled at Hassan.

Hassan, panting, looked at Klaus. Behind Hassan, Edouard was standing, and he heard it all. Klaus knew he had said something he wasn't supposed to say.

"Take it back," Edouard commanded his father.

"I am so sorry, please don't leave," Klaus requested Hassan to stay.

"I won't leave. I don't hide from anything. You said they are not my kids; guess what? You're mistaken, they are my kids, and I have every right to defend them. She is a single mom. Do you have any idea how it is hard for her? So, he is older than her, so are you. This is the eighth date, and yet he didn't get her pants off. He treats her like a lady. The first thing today, he asked her was about Peyton. You should be happy for her; she found a good father for her daughter. Do not stand in front of me and yell."

"I am sorry. I don't know why I said that," Klaus apologized to Hassan for misbehaving.

"You are a father. You have the right to defend their happiness," Hassan tried to motivate Klaus.

"Son, you should be happy to have such a wonderful guy who cares about your family. Obviously, his family too. And Hassan (long sigh), I don't know how to say this. You are a real parent for them, don't ever stop taking care of them," Valerie appreciated Hassan's efforts and tried to put some sense in Klaus.

"They are part of my world," Hassan said.

"Are you angry at me?" Klaus asked.

"No, I love you." Hassan smiled.

"What did you say?" Klaus asked Hassan to say those words again.

Klaus walked toward Hassan, panting. "Please say it again," Klaus said.

"I love you," Hassan confessed again.

"I was afraid to tell you that I love you, but you are in love with me too," Klaus was in tears.

"What are you waiting for? Kiss him, son." Denis stood up and said.

Klaus wrapped his arms around Hassan and kissed him. Then, Klaus put his forehead on, he smiled and said, "I love you."

"I love you too." Hassan replied.

"Now I have to ask you to move in with us." Klaus told Hassan.

"You want me to move here?" Hassan could not believe what Klaus had just said.

"It doesn't make any sense to live in another house. I am here; our kids are here," Klaus said.

"Our kids!" Hassan ratified Klaus.

"Yes, our kids," Klaus agreed.

"Klaus is right, you need to move here," Valerie insisted on Hassan too.

"Yes, please," Edouard followed his grandmother's lead.

"It is the right thing to do," Denis motivated Hassan as well.

"Okay," Hassan agreed.

"Okay, so you will move in with us?" Klaus asked Hassan to confirm his decision.

"Yes, I will move here," Hassan confirmed.

"Son, you've found a good man, don't lose him," Denis told Klaus.

"Next visit, I want a wedding," Valerie said excitingly.

"A wedding!" Klaus and Hassan screamed together.

"Why not? You are both deeply in love, and you share kids," Valerie said to both of them.

"If you both get married, remember, no divorce," Denis told the love birds.

"It's a one-way trip. Once you are in, you cannot leave. Are you in?" Klaus tried to make sure Hassan knew what he was doing.

"All the way long," Hassan approved.

Klaus took Hassan and gave him a big hug, followed by a passionate kiss. He closed his eyes then took a deep breath. Then, with a big smile on their faces, everybody looked at Hassan and Klaus.

Afterward, María lit the candle and began to pray. She saw Lizzie standing in her room after she finished praying. "I think it's time to move on," María said. Lizzie looked at María and said, "Not yet."

"Everything is going to be okay. Nicholaus and the kids are happy now," María said to Lizzie.

"The big dark cloud will come sometime soon. We are here for him and our beloved," Lizzie said.

"We!" María got confused by what Lizzie said.

"I am not alone. I brought some company," Lizzie clarified.

María saw a man and woman holding hands, a girl and a boy holding each other's hands. She couldn't see their face, and a young man stood up in the dark. María put her hand on her own, panting. "You are all here for him and his family?" María said.

"We are all his family, and you are also part of his family." With this, Hassan's mother came in.

"We are here for our son." Hassan's father joined the party too.

"Is he dying?" María asked as she was confused by what was happening.

"He is not dying. He will live long enough, but something terrible will happen to him," Lizzie told María.

"I am here with you," María was crying.

****

Hassan was preparing for the dinner with Cara and Pete on Saturday. Archie played with John and Carlos. Kyle spoke to Tristan, and Andrew spoke to Oliver. Ethan had to buy some things so he prepared a list and went to the supermarket with Cara and Pete. Hassan was concerned about how he would give everyone the news about Klaus. Cara noticed that Hassan was anxious, "Are you alright?" Cara asked.

"I don't know how to break the news to you all," Hassan told Cara.

"Tell me, I might help," Cara was hopeful that she could help Hassan.

"Don't tell anyone yet," Hassan warned Cara.

"Tell them what?" Cara asked.

"I am moving in with Klaus," Hassan broke the news to Cara about him moving in with Klaus.

"What?" Cara screamed.

"Is everything alright?" John asked.

"Hassan won't let me use the salt," Cara changed the topic and stalled John.

"You know him," John replied, unaware.

"It is time to move in with him," Hassan whispered to Cara.

Cara was happy that Hassan was moving in with Klaus but unhappy as they were all going to miss him, especially on the Friday nights and insisted Hassan visit them at least on Fridays.

"I never thought this day will come, but you have to come here to spend every Friday with us," Cara said to Hassan.

****

Chapter 19: Faith

"Everyone is a moon, and has a dark side which he never shows to anybody" - Mark Twain

Deep inside, Cara was not ready to let Hassan depart from their wonderful abode. Hassan had put in a lot of energy and he was the one who kept everyone intact, and now that he was leaving, Cara knew that it would be difficult for everyone.

"I am happy for you," Cara said to Hassan and hugged him.

"Are we hugging now?" Pete said, entering the kitchen.

"Hassan has good news. He will tell us later," Cara told Pete.

"Great, so we will celebrate," Pete said.

"Klaus and his kids are here," Kyle told everyone there.

Hassan opened the door, and Bella came running to play with Archie. "Oh, Archie has found a new friend," Hassan said.

"I missed you," Klaus said to Hassan as he was longing to see his beloved.

"Me too," Hassan shared mutual feelings.

"Hey, we are waiting here," Charley said, teasing them.

"I am sorry. Please come in," Hassan chuckled a little.

Klaus stood next to Hassan while everyone entered.

"Charley brought her boyfriend," said Klaus.

"That's okay."

"Did you tell them?" Klaus asked Hassan if he had told everyone about him moving in with Klaus.

"Not yet," Hassan denied.

"It's okay, you will find the perfect time," Klaus said.

"Can I see your paintings?" Edouard asked Hassan.

"Sure, come with me," Hassan was more than happy to show Edouard around.

****

Hassan and Edouard went downstairs to look at the artwork. While Cara was getting Peyton in her arms from Charley, Adrien was having fun with Bella and Archie. Kyle observed Cara's gaze as it shifted to Peyton. He saw how excited she was to see her. He thought may be Cara also wanted a child. In the meantime, Andrew and Klaus were discussing.

Klaus smiled after a while, and Andrew asked him, "Why do you have a creepy smile on your face?"

"Now I get it. Why Hassan calls all of you 'the gang'"

"That is rude, and so is Hassan," John didn't like Klaus' response.

"He told you this?" Ethan asked.

"Indeed, he did," Klaus replied.

"He is right. You all seem like a gang," Oliver agreed with Klaus.

"What else did he tell you?" Tristan asked Klaus.

"I am here," Hassan joined in.

"Do you have anything to tell us," Andrew instantly confronted Hassan.

"Yes," Hassan nodded.

"You are a traitor," Ethan did not like what Hassan told Klaus about them.

"Everyone attack him," Andrew ordered everyone.

Hassan was attacked by Andrew, Tristan, Ethan, John, Kyle, and Carlos, who held him and tickled him until he complied. And Hassan was giggling crazily while Klaus was beaming.

Hassan said, "I will move in with Klaus," Everyone stopped and looked at Hassan. Andrew was panting, "No, you won't," he said.

"Please…" Hassan was about to say something but got interrupted by Andrew.

"No, you won't. It starts with two nights, then three nights, and now he wants a full time with you. What about us?" Andrew asked in anger.

"He has found his happiness. And we cannot stand between him and his happiness," John regretfully said.

"Will you visit us?" Tristan asked Hassan.

"Every week, I will be here," Hassan replied.

"We will miss you," Carlos said to Hassan.

"I don't know what to say," Ethan did not have any words to express his sadness.

"You can say no," Andrew said to Ethan.

"There is nothing to say. He has the right," Kyle supported Hassan.

"No," Andrew denied Kyle.

Andrew ran to the yard and sat on the swing, and started crying incessantly. Hassan was looking at Andrew, and he did know what to do. While Klaus put his hand on Hassan's shoulder and said, "I will talk to him." Klaus went to talk to Andrew and sat next to him.

"What do you want?" Andrew misbehaved with Klaus a little.

"I know you and Hassan are very close but don't you want him to be happy?" Klaus asked.

"Please don't take him away from us."

"You all are his family. No one can take him away from you. But he is my family too. I love him, and I cannot breathe without him. Please don't stand in the way of our happiness."

"If you ever hurt him, I will kill you."

"I adore him, and I will never hurt him." Klaus took a long sigh before replying to Andrew.

"Can I come to visit anytime I want?" Andrew asked.

"Of course," Klaus replied to Andrew.

"And he will spend every Friday night with us?"

"Sure."

"Please make him happy."

"I will wait for few months, and I will ask him to marry me."

"You are a freak, and now I have to like you."

"Admit it. You like me."

"And why is that?"

"Because I make him happy."

"You are right. You do make him happy."

"So, do we have an agreement?"

"Yes, we do."

Andrew hugged Klaus, and then they entered home.

****

Hassan was standing, looking at Andrew with tears. He wiped Hassan's tears, "Move with him, be happy. Don't worry about us," Andrew said and hugged Hassan.

"Don't ask me to move in with you," Charley was in tears and said to Anton after watching everyone's reaction to the news of Hassan move in with her dad.

"At some point, you will," Anton replied.

"I am very happy," Adrien said.

"And why is that?" Tristan asked.

"My dads will live together," Adrien responded.

"Your dads?" Ethan asked Adrien.

"Yes, he is our dad too," Edouard was happy with Adrien as well.

"In this case, you have many uncles and one aunt," Kyle said to the kids.

"Wow, now we are a huge family," said John.

"Do I have to memorize all of your names?" Pete said with a smirk.

"Honey, you need to cook, not memorize," Ethan replied to Pete.

"I think John wants to tell us something," Tristan interrupted everyone and diverted everyone's attention to John.

"I haven't told him yet," John whispered to Tristan.

"What do you want to tell us?" Carlos asked John.

"Actually, I want to ask you to move in with me," John decided to get it off his chest.

"I don't know what to say," Carlos was overwhelmed with joy.

"Shut up and say yes. We all know that you are in love with John. John is a unique gem. Don't let him get away," Andrew smiled and said to Carlos.

"Say yes, and I promise you won't regret it," John sort of proposed to Carlos.

"I love you," Carlos confessed to John.

"I love you too," John took a long sigh and replied.

"I am a little sad, and you need to kiss me like a million kisses," Andrew flirted with Oliver.

"You are greedy," Oliver replied to Andrew.

"You like me," Andrew said to Oliver.

"So, I am the only one single again," Tristan said.

"Actually, someone wants to ask you on a date," Hassan said.

"Who is he?" Tristan asked.

"Remember, David? Our assistant chef?" said Cara.

"The cute guy!" Tristan exclaimed.

"Yes, he's always asking about you," Pete confirmed.

"Always!" Tristan remembered the guy like the back of his hand.

"He saw you once and kept asking about you. Apparently, he is into you," Hassan gave the exciting news to Tristan.

"I can give you his number," Pete said to Tristan.

"Isn't it a little weird?" Tristan asked.

"Do you want to be single for a long time? Trust me, you don't want that," Andrew said sarcastically as always.

"You know all of us, we all have the same mutual person, and he likes to be surrounded by quality people," Ethan told Tristan.

"Trust us on this. You won't regret it," Kyle assured Tristan.

"I guess I can give it a try," Tristan said.

"To be clear, Hassan somehow did match you all?" Charley tried to clear her confusion.

"Yeah, except Carlos and I," John responded.

"Interesting," Anton was intrigued.

"My boyfriend is a matchmaker!" Klaus said while smiling.

"No, I am not," Hassan denied it.

"Can I say something?" María wanted to share her opinion.

"Sure," Andrew answered.

"It is a good thing that you would be surrounded by the people you love. You don't give up on them, and they won't give up on you. The strong and deep connections will come out stronger after conflicts, as long as the conflict is something that can be resolved. And all I can see is love. You all have strong deep connections with each other, which uniquely makes unbreakable family roots. You all are not related, but somehow you all make a family. That is very rare," María's eyes held tears while saying this.

Hassan hugged María and kissed her forehead and said, "You are a part of our family."

Klaus looked at Hassan with a smile on his face, and all he was thinking was, "That's why I am deeply in love with you. You turned my world upside down. I can't wait to wake up every morning looking at your face. I love you."

Andrew took a long sigh and said to himself, "They all are a family for me, but you are the real family I always wanted."

Oliver was looking at everyone, thinking, "Is this how the real family looks like? I don't know what is the next step for me, but I have to live my life. My old life cannot be restored, but I am alive and still have the future. I have them all by my side, especially him."

****

Anton stood up, walked toward Cara, and asked her to give Peyton to him. He took Peyton in his arms. He took a deep sigh and said to himself, "I always wanted to have a baby, my wife and I tried for so long. I hope that I can be your father. I will love you and cherish you," He kissed her on the forehead.

Charley and Klaus were looking at Anton. Charley felt that Anton would be a wonderful father to her little girl. Klaus walked to Anton and saw Peyton holding Anton's finger with her little hand. "Do you feel the joy of holding a baby?" Klaus asked.

"I've never felt like this before. I have always wanted to have a baby," Anton replied with tears in his eyes.

"You know they say the babies know their real parents by touching and feeling. They cannot lie about them. Maybe you are not her biological father, but in her eyes, you are the real father," Klaus said to Anton.

"What do you mean?"

"Look at her. She is smiling, touching you. She likes you."

"She is breathtaking, just like her mother."

"Please, raise her well."

Anton looked at Klaus and started to cry, "Do you mean it?" Anton asked.

"I will never find a good father for her like you."

Anton looked at Peyton, and she smiled. "Hey, you. I am your dad," Anton said.

Hassan looked at Charley and whispered, "He will be a good father."

"He is a good guy," Charley agreed with Hassan.

"Can I tell you something?" Hassan asked Charley.

"Sure."

"Share Peyton's parenting with him."

"You think so?"

"Did you look at him with Peyton? He is her father; you knew that already in of your heart."

"We just started…."

"Look at her and look at him. This won't happen every day. If he is looking for the best for her, don't take her away from him."

"I don't know what to do!"

"A mother always knows what's the best for her baby. Please, tell me I didn't have a big fight with your father for nothing."

"I'll do my best for my baby," Charley replied to Hassan with a smile.

Hassan hugged Charley and kissed her on the head, "That's my girl!"

****

"So, when will you move in with me?" John asked Carlos.

"Maybe next year," Carlos teased John a little.

"Next year!" John was shocked.

"I am kidding. I will move in with you next week. Is that good for you?" Carlos laughed.

"Yeah, sounds good to me," said John while biting his lip.

"When will we eat?" Pete asked.

"All he ever thinks about is food!" said Andrew.

"Mind your own business," Ethan retorted to Andrew.

"He wants you to try the new dish that he prepared," Hassan told everyone about Pete's attempt at a new dish.

"I am hungry," Oliver was famished.

"I bet you are," Andrew said to Oliver and winked.

"How do you manage to make everything sound dirty?" Ethan was confused about Andrew's expressions.

"Maybe you can take some pointers from him," Pete said.

"That is tough," John laughed.

"Do I need to remind you all my kids are sitting here?" Hassan diverted the topic.

"Sorry, won't happen again," Andrew apologized.

"I am hungry," Adrien said.

"We will set up the table now, then we will eat. Everyone let us set up the table," Hassan called everyone.

"Your dad is really scary," Andrew whispered to Adrien.

****

Klaus was watching Hassan with the others, setting up the table. Edouard looked at his father and said, "Stop staring at him like that."

Klaus took a deep sigh and said, "I want to marry him."

"Dad will move in with us. I am so excited," Adrien was already excited about Hassan moving in with them.

"We are all excited," Charley felt the same as Adrien.

"How are things going between you and Anton?" Finally, Klaus asked Charley about her ongoing relationship.

"Things are going great."

"I am happy for you."

"Did you have a fight because of me?"

"Yes, he did, and he said awful things to him," Edouard told Charley about Klaus' fight.

"What kind of things did he say?" Charley asked.

"I said terrible things, but here we are. He fought for you, and he proved his point. He is a real parent for you all," Klaus explained.

"Dinner has been served!" Hassan said loudly to invite everyone to the dinner table.

Dinner was served at the large table, and everyone sat around it. While Anton was holding Peyton, Hassan took Adam and fed him. Pete was pleased to see that everyone seemed to appreciate the cuisine. Anton smiled as he stared at Charley. Pete was watching Ethan eat while Carlos was feeding John. María smiled as she glanced at Nicholaus, Charley, Edouard, and Adrien, and she thought to herself, "You all haven't sat at the same table for years until he came into your life, and now you are sitting every day at the same table." Andrew grinned as he stared at Oliver. Then, with a bashful smile on his lips, Oliver stared back at Andrew.

Edouard paused for a moment, wishing his mother was present. María noticed Lizzie putting her hand on Edouard's shoulder as she gazed at him. Edouard sensed his mother's presence. He smiled as he took a deep breath. Hassan fixed his gaze on Adam and thought to himself, "Dad, I wish you were here to meet your grandkids. You would have loved them."

"I am really proud of you, son," María remarked as she glanced at Hassan and saw his father caressing his head. Hassan inhaled deeply, closed his eyes for a few seconds, and then opened them again. He had the impression that time had frozen and that no one was moving. He got out of his chair and circled the table. He smiled as he looked around at everyone. He noticed his father, who was smiling, standing next to his seat... Hassan told his father, "I miss you." Hassan's father grinned and flung his arms open. Hassan ran to hug his father.

"Look at you, feeding your own child. You had a striking resemblance to me. I used to nurse you when you were a baby by putting you on my lap. I am extremely proud of the person you have evolved into," Hassan's father said.

Hassan grinned and closed his eyes as he gazed in the mirror. Then, Hassan inhaled deeply and opened his eyes.

"Is everything alright?" Klaus asked Hassan.

"Everything is perfect," Hassan stalled.

Klaus held Hassan's hand and said, "I love you." Hassan smiled and replied, "I love you too."

They sat together after dinner and had some conversation. It was time to go home then. Klaus had taken the children and María while Anton had driven Charley and Oliver. Andrew joined Hassan in his chamber, who lay on the bed beside and read a book. Andrew took Hassan's book, closed it, and put it to one side. Andrew rested his head on Hassan's chest and pleaded with tears, "Don't forget us, please."

"You are all my family; how can I forget you?" Hassan guaranteed Andrew.

"You are my closest friend," Andrew said.

"I will be here every week," Hassan replied.

Andrew looked at Hassan and said, "You will sleep between my arms till you move in with Klaus."

"Why do I always have to sleep between your arms? Can't you sleep between mine?" Hassan asked Andrew.

"No offense, your arms are small, and you are tiny for me."

"How dare you."

"Shut up already and go to sleep."

Andrew wrapped his arms around Hassan and kissed him on the forehead. "Klaus would kill me if he sees us now," Andrew said.

"Why would he?"

"No kidding, he would because you are his heaven on earth. No one wants to see his heaven like this."

"He knows we are very close."

"Try to sleep."

 "Good night."

"Good night."

****

Hassan closed his eyes, and there began his darkest nightmare. In a huge structure full of sharp items on the floor, he saw himself wandering alone. He floated at a location devoid of sharp items. He paused. He was frightened. He spotted a man seated on a chair wearing his face cap. "Are you not unbeatable?" questioned the man. The man then walked, and suddenly, the pointy objects began attacking Hassan. He raced as quickly as he could, and sharp objects pursued him. He had just gone to an impasse. He stood weeping before the wall. He closed his eyes and noticed the sharp objects approaching him swiftly. Just before that, he opened his eyes. He breathed deeply, and he felt something move on his body. He couldn't make a sound or even move. The chain was moving around his torso, his neck, his legs, and his arms. The chain thrust him through the wall and he fell from the cliff. While he was still falling, he closed his eyes. He felt that he would fall forever. His body shook within that experience. It was one of the worst feelings he had ever experienced. When his body hit the water, he opened his eyes. He could not breathe and Hassan continued to drown as the chain drew him deeper. He could not breathe.

He saw Brian was crying and saying, "I am sorry." "I want to wake up. I want to wake up." was all Hassan's mind could think. Hassan felt someone shaking his body, screaming, "Wake up!!" Hassan finally opened his eyes, and he was gasping for air.

"Are you okay?" Andrew asked a frightened Hassan.

"It was just a nightmare," Hassan replied while panting.

"The same old nightmare!"

"No, this was something different."

"It's just a nightmare."

"Not to make it sound a drama, but this one was terrifying."

Andrew hugged Hassan and comforted him, "It's just a bad dream."

"It wasn't just a bad dream."

"What do you mean?"

"I was facing my own fears, and I couldn't."

"Try to sleep. I am here."

Finally, Hassan fell asleep between Andrew's arms and held Andrew's t-shirt like a baby with his hands. Andrew recalled what Nie told him and his tears began to fall tear after tear from his eyes. "I wish I could save you, but I don't know how," Andrew kissed Hassan on his forehead.

****

Andrew went to Klaus' residence a few days later. Hassan was sad to leave his home, and every single day, he would feel the same as he did before. He knew that he wouldn't be able to see Andrew, Cara, Pete, Ethan, John, Kyle, Tristan, Carlos, or Archie.

But he was also happy to live with Klaus, the one he love, and be with his kids. He was torn between his emotions. Andrew stopped before Klaus' house, "Are you ready?" Andrew said. He looked at Hassan; Hassan was crying.

"I don't want to lose you all," Hassan sobbed.

"Don't be silly. You will come to visit us every week. You won't lose us. So go, be happy with him, and if he ever hurt you, I will hunt him down," Andrew bucked Hassan up.

"Thank you," Hassan said to Andrew.

Andrew had driven Hassan to Klaus' house, and everyone was waiting for him at the entrance.

"They love you. We all love you. Don't make them wait," Andrew said to Hassan.

Hassan opened the car door and went to them. He ran to Klaus, who was smiling. Klaus placed his arm securely over Hassan's body and kissed him. Andrew looked at Hassan's face with a large smile. Klaus walked down to Hassan and brought the kids to see him.

"Hey, Andrew. How are you doing?" Klaus asked.

"I am doing well. Thank you for asking. How are you doing?" Andrew asked back.

"I am very happy and excited. He is moving in with me," Klaus replied.

"Make him happy," said Andrew.

"He is my heaven; I would do anything for him."

"You are his heaven too. You gave him everything he wanted. You were a one-night stand, but now...."

"Now what?"

"I can't believe what I am about to say. Your goal was casual sex, but here we are, standing in front of each other."

"And your point is?"

"I am jealous of you."

"Are you in love with him?"

"No, but he is my real family. Please take care of him."

"He is not going anywhere. He is your family too, and no one ever can take your place. I promise you I will take care of him."

"I know that."

"Come in, have lunch with us."

Andrew walked into Klaus' house and stayed for lunch, and then he left after a few hours.

"Let us arrange your stuff," Klaus said to Hassan.

"We already did it," Hassan told Klaus.

"When?" Klaus got startled.

"You were busy talking to Andrew while we did everything," Edouard said to Klaus.

"Your boyfriend is very efficient," Charley said to her dad.

"What did we agree on?" Klaus asked Charley.

"I feel weird," Charley replied.

"What did you agree on?" Hassan was curious.

"To call you dad," Charley said to Hassan with a big smile on her face.

"It is not something you can force them for. They have to say it naturally. If she is not comfortable saying it, don't push her or anyone to say it," Hassan calmly told Klaus.

"I am sorry. I just want them to see you as a parent," Klaus apologized to Hassan.

"This is not how it works. If I am a parent, I have to earn it. Parenting is an action, not a word you can say," Hassan smiled.

"I will always call you, dad," Adrien said to Hassan.

"I would like that too," Edouard agreed on doing the same.

"For me, it is a little weird to say it, but you are a real parent for us. in fact, we have three parents," Charley said.

"See, you don't need to push them to do something," said Hassan.

Klaus smiled then said, "I have nothing to say." Hassan walked to Klaus, stood up on his toes, and kissed him.

"I just want to let you know; tomorrow Nina will visit us with her wife," Klaus informed Hassan about Nina and her wife's visit to their place the other day.

"And?" Hassan asked.

"They want to talk to us," Klaus answered.

"About?"

"No idea."

"Don't worry. We will take care of them."

"I love you."

"I love you too." Hassan smiled.

The next day, Klaus was worried about Evelyn (Nina's wife) meeting Hassan. Nina noticed that Klaus was worried. Regardless of that, Hassan, along with the rest of the family, started prepping up for Nina and her wife's visit to their home.

****

Chapter 20: Stalker

"Life has many ways of testing a person's will, either by having nothing happen at all or by having everything happen all at once" - Paulo Coelho

After Hassan's shift to Klaus' house, Nina and Evelyn visited Klaus, Hassan, and their kids. They wanted to talk to Hassan and Klaus in private.

"We decided to have a baby, and we want Hassan to be the donor," Nina said.

"What?" Hassan could not absorb the news at that moment.

"We want you to be the father of our baby. If you agree, of course." Evelyn reiterated what Nina had said to Hassan.

"Why?" Hassan asked.

"You are smart, sexy, and we want our kid to grow up to be like you," Nina replied.

"In normal circumstances, I would say yes, but I cannot, sorry," Hassan refused the ladies.

"Can you tell me why?" Evelyn asked.

Hassan looked at Klaus. Klaus held Hassan's hand and said, "It's okay."

"What is going on?" Nina was confused about the situation.

"Hassan has been diagnosed with HIV," Klaus disclosed the news about Hassan's health.

"How did he get it?" Nina wanted to probe a little more about Hassan's disease.

"From his ex," Klaus replied.

"Does he have an undetectable viral load?" Evelyn asked.

"Yes," Klaus replied.

"I don't see a problem then. Hassan, I am a doctor, and you don't need to be worried. You just gave us another reason to have your baby," Evelyn was readier than before.

"Hassan, look at me," Nina said to Hassan.

Hassan looked at Nina, and tears were forming in his eyes. She wiped his tears, "I don't want to see tears in your eyes," Nina said to Hassan.

"It's settled then. I will come by tomorrow to take you to the fertility center to do some tests. Then we will take it from there. I hope our baby will have your eyes," Evelyn was excited.

"Ladies, there is only one Hassan, and he is mine," Klaus showed his possessiveness for Hassan.

"Don't be greedy; give us some," Nina said.

"He is all mine," Klaus responded to Nina.

"And Klaus, no sex for a week," Evelyn forbade Klaus from being too intimate with Hassan.

"A week!" Klaus was certainly not ready for that news.

"We need him to save some for us," Evelyn said to Klaus.

"Are you okay with this?" Klaus asked Hassan.

"We can touch, kiss and more," Hassan tried to offer Klaus some flexibility in intimacy.

"That will be torture," Klaus disagreed.

"You can wait, and I will do anything you want me to do," Hassan assured Klaus.

"Anything! Are you trying to seduce me?" Klaus said to Hassan.

"Just to tell you what is waiting for you," Hassan said.

"Okay, ladies, we have a deal," Klaus agreed with the ladies.

"Klaus, no offense. You are like a teenager," Evelyn said, Klaus.

"We've never seen you like this," Nina agreed with Evelyn.

"Hassan, can you check on the kids?" Klaus asked Hassan.

"Okay," Hassan said and got up to leave.

Hassan kissed Klaus and left the room. Klaus looked at Evelyn and Nina and started sharing his views about Hassan.

"He is different from anyone I've ever met. He brings out the best in me, and I cannot lose him. He is my heaven. I know who I am deep down, and it took a long time for me to love myself, but I finally do. He entered my life and turned everything he touched into something extraordinary. I know what I bring to the table and what I'm capable of when I'm in love, but everything is different with him. Have you ever seen my kids this happy? The answer is no! He chose to be there for them, not to replace her. He repaired my broken family. So, excuse me if I am acting like a teenager with him and feed that hunger, I have had for years now."

"That's why we chose him to be the father of our baby, and we want him to be involved in the baby's life," Evelyn said to Klaus.

"Are you planning to marry him?" Nina asked Klaus.

"I planned to propose to him after few months." Klaus replied.

"A few months!" Nina exclaimed.

"I cannot wait for more. I want him to have my name," Klaus instead preferred to marry Hassan quickly.

"I am happy for you," Evelyn said.

"I told you. She would love him," Nina said to Klaus.

"I want to know more about him," Evelyn was eager to know Hassan more.

"He is brilliant, naturally a parent, wise, a gifted artist, and I love him," Klaus attributed to Hassan.

"What do you mean by a gifted artist?" Evelyn asked.

"You should see his pictures and paintings," Klaus told Evelyn about Hassan's artwork.

"I'd like that," Evelyn said.

"You need to know that your baby will have three brothers and one sister," Klaus told Nina and Evelyn.

"We are counting on it," said Nina.

****

Two months later, Hassan woke up to a kiss from Klaus. He opened his eyes to see Klaus smiling. He smiled and kissed Klaus back.

"Good morning," Hassan greeted Klaus.

"Good morning," Klaus greeted Hassan back.

"Why are you up early?" Hassan asked.

"I am just excited," Klaus replied.

"I can see that,"

"Will you love me even if I am very old? And would we have old people sex?"

"What a question? I will love you, no matter what. I choose you for the rest of my life."

"I love you."

"Is there anything you want to tell me?"

"Tonight, at dinner, I have something to tell you."

"Okay, then. Stop looking at me like that."

"Like what?"

"Like I am your breakfast."

"Can I get some?" Klaus bit his lip.

"I am all yours," Hassan smiled.

Klaus and Hassan shared long passionate kisses. At that time, Bella entered the room with Adrien. She jumped on the bed and started to lick Hassan's face. "Hey, he is mine," Klaus said.

"We are waiting for you downstairs," Adrien said.

"Everyone is up early! What is going on?"

"We decided that today is for family," Klaus told Hassan.

"I will pretend to believe you," Hassan could not believe Klaus.

"Trust me," Klaus said.

"I love you. Adrien, we will be there soon," Hassan was happy.

Klaus and Hassan washed their faces and brushed their teeth. Klaus went downstairs, Hassan stood in front of the mirror, and he looked at himself. What he was feeling, he could not understand. He was shuddered by mixed feelings. A warm tear fell from his eye. "What is happening with me?" Hassan was confused. He washed his face and walked downstairs anyway.

"Good morning, everyone!" Hassan greeted everyone downstairs.

"Good morning!" Everyone replied to Hassan.

María was preparing the juice. She turned around and looked at Hassan, she suddenly dropped the liquid on the ground, and the juice pot was broken into pieces.

"Are you okay?" Hassan asked María.

"I am okay. I will clean it," María replied.

"No, have a seat. I will clean it. Nobody moves," Hassan said.

Hassan began cleaning, and María sat on a chair, watching Hassan. She and the others around him saw Lizzie. It was the day, she knew. But she couldn't do anything, and she wanted to stop what were to happen next. She felt somebody tightly squish her heart. She didn't know what she should do. Edouard felt his mother's presence, but his feelings could not be understood. While Hassan cleaned up, Peyton started crying all of a sudden, "I will take care of her," said Hassan.

"You are a saint," said Charley to Hassan.

Hassan took Peyton between his hands and fed her.

Charley looked at her father, "You look weird today!" Charley said.

"Tonight, I will ask him to marry me," Klaus told everyone.

"Oh, my God. I am so happy for you," Charley said.

"Try to keep it down," Klaus told her to lower it down so that Hassan could not hear it.

"It's a happy day," Edouard said in joy.

"Yes!" Adrien was joyous as well.

"María, do you have anything to say?" Klaus asked María.

"You both look perfect together, and I am happy for you." Finally, María gave her well wishes to Klaus.

"He doesn't know, and I'd like to keep it that way," Klaus told María.

"We will make sure of it," Charley assured her father.

"I need help from all of you. Also, the gang will be here today," Klaus informed everyone about the arrival of Hassan's friends.

"Stop calling them that," Charley was not happy with her father's remarks.

"That's what Hassan calls them," Klaus justified.

"I agree with him," Edouard supported his father.

"He is coming. Act natural," Klaus said.

"I've fed her and put her to sleep," Hassan said to Charley.

"You are a saint," Charley praised Hassan once again.

"You are welcome. So, what do you want to do about dinner?" Hassan asked Klaus.

"We planned everything. We just need you to be there," Klaus answered Hassan.

"Okay!" Hassan nodded.

"The whole gang will be here," Charley used the same word for Hassan's friends as her father did.

"Also, Anton, Nina, Evelyn and my parents," Klaus informed every one of the guests for dinner.

"Grandma and grandpa will be here too?" Adrien was excited to hear about his grandparents coming over for dinner.

"Is there anything you need to tell me?" Hassan asked.

"It will be a family dinner, a celebration of our big family," Edouard said with a big smile on his face.

"You mean, our huge family," Charley rectified Edouard.

"I've always wanted a big family," Hassan said

"Your wish came true," said Klaus.

"Indeed," Hassan agreed with Klaus.

"Have a seat and eat something," Klaus said to Hassan.

"Okay. After breakfast, I will go to the restaurant to check on how things are there, and I have to go to the pharmacy to pick up my meds," Hassan said.

"Take your time," Klaus smiled.

"You look weird today," Hassan looked at Klaus strangely.

"I am happy," Klaus justified his expression.

After Hassan left, Klaus started preparing all his plans with the help of the children. He also asked Cara, Andrew, and Ethan for assistance.

****

Hassan worked for hours at the restaurant. The time passed so quickly, and he didn't notice. Klaus called him to tell him that they were all expecting him. Hassan replied, "I have to go to the pharmacy to pick up my meds."

Hassan quickly left for the pharmacy. On his way, he felt being watched by someone. He entered the pharmacy, collected his medicines, and left. He felt someone looking at him every minute. He continued to walk, but he was panting. He wasn't able to understand why he was. Finally, he stopped and turned, but he saw something that made him realize why he got that feeling. He shut his eyes, and it was all dark he saw afterward.

***

"Hassan said he'd go to the pharmacy two hours ago," Klaus was worried, broken, and wept as he was concerned.

"Did anyone of you tell him?" Klaus said.

"No one told him anything," Andrew denied.

"He will be here soon," John was sure of Hassan's return.

"Can anyone call him?" Valerie asked.

Klaus called Hassan, and someone answered.

"Who am I talking to?" Klaus asked.

"I am Jason, an officer of Toronto PD. Do you know the phone owner?" A police officer asked Klaus.

"I am his fiancé. What is happening? Is he okay?" Klaus was starting to panic.

As the officer talked to him, everybody looked at Klaus. He sighed for a long time, and tears started falling from his eyes. He threw the phone in his hand. Andrew took Klaus' telephone and spoke with the policeman. Andrew began to cry after he ended the call.

"Someone tell us what is going on?" Cara was anxious.

"Hassan has been kidnapped, and the people who saw the incident reported it to the police. They found his phone," Andrew broke the news to everyone.

"Are you serious?" Ethan could not believe what he heard.

"He is gone," Klaus started crying.

"He won't leave us. He will get back to us," Edouard started crying as well.

Valérie hugged Edouard and said, "He will."

Adrien was also crying, so Cara hugged him. Everyone was under the shock of the news.

"But why? Who could have done it?" Pete sobbed.

Everybody cried. Nobody had seen it coming. What was supposed to be a good day had taken an unexpected turn. Ten minutes ago, the house was full of delight and joy, but suddenly, sorrow and tears took over. Nobody said a word for a while. Klaus' heart was broken, but there was no way out of it this time. There was never going to be the same family again.

He walked to Andrew and yelled, "I want him back." Andrew wiped his tears and said, "We will look everywhere for him. We will find him."

Andrew looked at everyone and yelled, "What are you waiting for? We need to start searching for him."

"We are all ready to do anything for him," John said.

"We need to divide into groups, and each group will search in a place," Nina told everyone.

"Valérie, Denis, Evelyn and Charley, you will stay here with the kids," Ethan said.

"I will go with Nicholaus. Cara with Kyle, Pete with Ethan, Tristan with David, John with Carlos and Nina with Anton. We will meet here after hours. I will never rest till I find him."

****

Andrew went with Klaus to the police station. Klaus was unable to stop his tears and control his sentiments. He did not know what to do.

"We have to find him as soon as possible; otherwise, we will lose him forever," Andrew said to Klaus.

"Don't you think I already know that," Klaus said in anger.

"You don't understand. If he doesn't take his medication, that is the end of him," Andrew replied.

"Why is this happening to us?" Klaus was furious.

"We have to put his picture on every news later, social media and everywhere," Andrew tried to maintain his composure.

After hours of searching, they came back home. They found nothing.

"Nicholaus, I used your laptop to do something. I used all the pictures you have of Hassan on your OneDrive," Evelyn had come up with a solution while everyone was busy searching for Hassan.

"What?" Klaus didn't get what Evelyn said.

"I created a website to help find Hassan. I posted all pictures of you with him. We have many people who will support you. But now you have to talk to them through a video and explain your feelings to them," Evelyn said.

"That is a good idea," Andrew liked Evelyn's effort.

"It's like having an army searching for him with us," Ethan said.

"We need to find him, and I am with you," So Pete joined the whole team to find Hassan.

"We will use everything we have to find him. I will not stay here doing nothing. He is my dad," Charley was crying.

"We will find him," Edouard could not hold his tears.

"How can I help?" María asked.

"You take care of the kids for him," Andrew said to María.

"I don't know what to say," Klaus was speechless.

"He is not just a friend. He is family," said Tristan.

"He always says, 'We don't give up on family.' So, we won't give up on him," Andrew said.

"We need to find him before it is too late," Cara said to everyone.

"What do you mean?" Kyle asked.

"Without his med, that could weaken his immune system, and he could become sick," Carlos said.

"We will find him," Valerie remained positive.

"Son, we are all with you," Denis assured his son.

"We will find him. It's just a big dark cloud, and it will pass," Klaus was optimistic about finding Hassan back too.

"What did you say? A dark cloud?" Andrew was panting.

"That's what Lizzie told me," Klaus said.

"Who is Lizzie?" Unfortunately, Andrew was not acquainted with Klaus' life.

"My wife," Klaus said to Andrew.

"Your dead wife told you about the dark cloud?" Andrew asked Klaus.

"Are you alright?" Cara asked.

"He will come back, but he will not be himself," Andrew said.

"What do you mean?" Ethan was confused about what Andrew said.

"Hassan had a dream about his father telling him about a dark cloud will come and destroy him, and he won't be himself until..." Andrew told everyone in detail about what happened.

"Until what?" Klaus asked.

"Until he will give up on himself," Andrew replied to Klaus.

"And what will happen next?" Charley asked.

"He didn't tell me," Andrew was clueless about the way forward.

"He will back to normal. He is not alone; Lizzie and his family are with him," María joined the discussion.

****

Everybody left after an hour, and Klaus kept speaking with people on Evelyn's website. He talked to a thousand people. Finally, Klaus shut down his computer and went to his room. He saw Hassan waiting for him when he opened the door.

"You are here. I was worried about you," Klaus said. Klaus held Hassan's hand. When Klaus saw Hassan wearing pajamas, he smiled. He walked between the candles, "I..." Hassan was not finished as Klaus held his hand.

In complete silence, Hassan and Klaus began to dance. Klaus looked at Hassan's face. They were swinging and turning. Klaus forgot everything, for he was in Hassan's arms. All he ever thought was, "I am in heaven.". Klaus smiled and turned around. He stopped when he saw himself in the mirror.

"This isn't real," Klaus said to himself.

He fell on his knees, looked at the sky through the window, "God, if you can hear me. Please return him to me. To us, to his family. I won't ask for much, just return him to me, please. I can't live without him. I don't know how to live without him. Please, God," Klaus said.

****

Hassan began waking up. But he could not see. He realized that he couldn't see despite opening his eyes, that's when he felt a blindfold. He couldn't speak or move his mouth because his mouth was like a duck type.

On his neck, arm, body, and legs, he felt a cold metal chain. He was chained to a pole behind his back, but the bar could move, and his feet had no ground. He felt fresh air flowing beneath his feet. He was scoffing, "This couldn't. No, No!" Hassan told himself. He felt something in his ears, but he couldn't figure it out. Somebody spoke suddenly and Hassan immediately recognized it was a voice changer tool.

The voice: "Hi, Hassan. You are awake now, and you are about to face your fears one by one. You are alone, and no one is here for you. You need to do it all by yourself. You are chained to a wooden pole at a specific height. You might think I am insane, but I am a genius. Now you need to stay in this situation for the whole night. Don't try to move much, or you will fall. If you care about your family and Klaus, you need to stay alive. And one more thing, I gave you your daily medication as you are supposed to, and I will give it to you every day at the same time. This is because I have a heart, and I want you alive."

Hassan was freaking out and attempting to move his body randomly. But he couldn't make a sound. He tried to cry and wept. Finally, his eyes were closed, and images of Klaus were seen. "This will kill me, and I cannot stand again. Is this written on my forehead like death? I don't know what to do, but this can't be my end. Klaus, I love you, and I am sorry if I couldn't make it." Hassan remained up for hours and tried to relax, but he couldn't. For a while, he closed his eyes, waking up to fall in fear. Three of his fears together was a living hell for him to live. What was this type of mind game?

Something he recalled. "This is what happened with Oliver. Is it possible? How did he find me? Is Brian with him? I don't know what to believe right now, but all I know is that I am the next victim of a mad man. I have to choose to live with my fears for the rest of my life or die here," Hassan told himself.

There were thousands of thoughts, but he couldn't focus on anything right now. He was scared, dying on his skin. Since he was only in his underwear, he was cold, but he could feel himself sweating. He felt the sweat drops moving down to his legs. He wanted to hear where he was, but he couldn't hear anything because of the airpods and tape. He fell asleep as his thoughts crowded his mind. Then, as he thought he had fallen into the water, he woke up frightened.

The voice: "Relax, it's just water to wake you up.

Hassan could feel his feet on the ground. He was happy.

The voice: "Now, it's time to take a shower. Be a good boy and don't try to do anything stupid, or you will find the people you love next to you.

Someone was removing the blindfold and as Hassan was finally able to see, he saw Brian in front of him. Hassan was panting. Brian removed the tape from Hassan's mouth. Hassan could not believe what he was seeing. Brian was chaining him to the washroom now. The wall chains were stabilized so Hassan could not move. Brian hugged Hassan from the back and wept.

"I don't understand," Hassan said to himself. Brian began washing Hassan's body. Peculiarly, Brian moved his hand on Hassan's back. Hassan couldn't tell what Brian was trying to do. Hassan shut his eyes and focused on his body. He was weeping hysterically now. "Let me just go," shouted Hassan.

The voice: Who do you want to be next to you? Pete or Andrew?

"I will do whatever you want me to do," Hassan was in tears.

The voice: "Good boy. You need to eat to keep your energy."

Hassan did not have any choice but to comply with the will of the mad man. His veins were in immeasurable pain; he could feel his body burning. He was breathing heavily. As Brian kept crying, Hassan walked away. Hassan looked down at his legs and saw the chain. As he looked in front of him, he saw a faceless person chained and the man next to him was wearing all black; he put the sword on the chained man's neck. Brian stopped to stable the chain in hooks on the wall.

Brian started to feed Hassan, but Hassan's eyes were fixed on the faceless man and the executioner with a sword on the faceless man's neck.

Hassan had no choice but to swallow his food; tears were falling like a waterfall from Hassan's eyes. The executioner moved his sword and separated the faceless man's head from his body. Hassan stopped chewing his food, looked at Brian with his eyes full of tears

Brian wiped Hassan's tears, put his head on Hassan's chest and cried. Hassan's tears didn't stop. Brian fell on his knees and wrapped his hands around Hassan's body and cried. Hassan saw the faceless man's head rolling on the ground. He looked at the head, and he could picture his face. Hassan closed his eyes, then opened them again to see his wrecked house and his sister's hand between the rubble.

Hassan couldn't live with the excruciating pain of losing everything once again. Yet, his world was crumbling in front of his eyes, and he could do nothing to stop it. This time, he couldn't find the strength to go through the same pain. He took a deep long sigh and wished his heart could stop beating so that everything would finally end. He felt his eyes closing, and he wasn't able to open them. "It could be the end. Well, I am too tired to carry on with all that pain," Hassan said and then he fainted.

****

Early on, María woke up, lit a candle, and began to pray to God for Hassan, his protection and safety. She took Hassan's photograph and held it close to her heart. Her tears fell silently out of her eyes. She saw Hassan's pictures floating in the air. With a sad smile, she looked at every picture. She went through all those images and tried to reach him, but she could not. Finally, she stopped spinning and took a long sigh, holding his picture close to her heart, closing her eyes. She then opened her eyes, kissed Hassan's photo, and placed it on the bedside table. The door was gone, and the light was turned off.

****

Throughout the night, Klaus couldn't sleep. He was swiping through photos of him and Hassan on his phone. Hassan was smiling in all the pictures. Finally, he stopped on his favorite shot. He zoomed in on it and kissed the picture. A river of tears started flowing down Klaus' face.

He placed his phone on his heart. He had a vivid memory of the first time Hassan said "I love you" to him. He looked up and walked toward the window. He laid his head over the window, and cried tears over his joyful memories and endless love. He felt that incredible feeling and somebody's presence in the room. Lizzie stood behind him and held his back while Hassan's father put a hand on his shoulder. "I want him back. I can't live without him," Klaus said.

****

Andrew looked at one of the paintings of Hassan. In that painting, Hassan had covered his eye with his arm. He always hated that painting because he considered it too strange. With one hand, he held out the painting and, with the other hand, touched it. His tears went free from his eyes. He went back to his room and sat on the table next to his bed, where he put Hassan's painting. Then he watched Mike's painting around the room.

He held the painting in his hands, and he spoke to it, "He's not here. He needs your help. He needs your assistance. All of us need your assistance. I don't know if I can move on. He's my family." Pete and Ethan entered the room as Andrew had left the door open. With tears in his eyes, Pete hugged Andrew. Andrew was held back by Ethan. There were no words, only the sound of their silence and desperate tears as they prayed for Hassan's safe return.

Overnight, Cara and Kyle escaped the sorrow by sex. They used sex to run the unbearable pain and the burning fire through their veins. When they were done, Cara put her head on the pillow, looked up at the ceiling, and then collapsed: "He is gone." Kyle said while holding Cara's hand. "He's gone," with tears.

Cara said, "I can't breathe."

Kyle held Cara between his arms.

***

Tristan looked at his dad's image on the stand next to his bed, and his pillow was wet by his tears. He got up and went to the window and saw Hassan running in the courtyard, laughing. Then he fell, and Tristan sat on his legs to prevent him from moving.

"Give me my phone," Tristan had said.

"Not before you tell me everything, and I mean everything," Hassan said. Tristan looked at the sky with eyes full of tears said, "I can't lose them both. One hit was enough for pain; another one will end me."

Adrien hugged Bella and couldn't move from his bed. "Pray with me that my dad comes home. I miss him a lot."

The voice: "Are you ready?"

Hassan did not even try to move. Hassan saw Brian silently crying. Brian had bent his blindfold and moved Hassan so that he could stabilize chains on the hooks of the wall. The scent of men stuck in Hassan's head. This perfume was a sign of suffering, torture, tears, fear, and obscurity. He felt that the darkness was near him and that it was about to touch him. Hassan felt something cold, like metal and sharp on his back. It was a sharp object, he realized. He tried to move and shout, but he couldn't. He began to move his body to prevent the intense feeling of that object. Hassan's skin was cut. He bled, but his fears had frozen his body. In a desperate effort to gain his stolen freedom, he tried to release himself. He stopped when he heard his right shoulder pop. He felt the pain burning in his shoulder. The worst pain is when your freedom is stolen, and the people you love are taken from you. At that moment, Hassan wanted his heart to be destroyed and finish it all. He took a long sigh and flew with wounded wings. The darkness suddenly stopped, and felt Brian's hands on his body. He tried to treat Hassan's wounds but couldn't. Then Hassan was still hanging in the air. He was exhausted and in pain. Soon, he felt like he was being moved down from the hanging position. Brian opened the blindfold and removed the tape when he reached the ground. "It's time for your meds," said Brian. "I don't want to take it. Let me die," said Hassan in a voice full of pain

"He is not here, but he has cameras everywhere. You need to take your meds not only for yourself but for Klaus. I promise you I will find a way to save you." Brian assured Hassan.

"He will kill you," Hassan replied to Brian.

"I was dead a long time ago, and I can't forgive myself for what I have done to you. So, please take your meds," Brian was in tears.

Hassan took his meds. Brian hugged him, "I love you, and I know you still have the amazonite I gave you. Please keep it to remember me after I am gone, and take care of my sister," Brian said.

"I can't," Hassan said to Brian.

"No, you can. You are Hassan," Brian was confident of Hassan.

Brian kissed Hassan and wiped his tears.

"Every time I touch you, I cause you more pain. Every time I touch you and your will has been taken away from you. If I could go back in time, I would do everything differently. Now I have to save you, but you have to fight."

"I can't. This is my last fall."

"You are a freaking warrior. No one can defeat you. I will save you, but I need more time. I will find a way to save you, so you will return to Klaus and your kids and family. Please don't break them. And if my sister had a child, don't let him become me. Love him like he is your own kid."

"What about you?"

"What about me? Don't worry about me; I will finally be free from all that pain. It would be enough if you remembered me."

"I will always remember you."

"One more thing, please forgive me. I hurt you so much. And I stole one of your paintings."

"I already forgave you."

"I hope you don't mind kissing you." Brian kissed Hassan on his lips, then took a long sigh. "I love you," Brian said and walked away.

Klaus didn't leave his room, eat or rest. So Valérie and Denis went to his room to check on him.

"You need to eat," Valerie tried to convince Klaus.

"I have no appetite," Klaus denied.

"You need to keep your strength," Denis tried to buck up his son.

"I want him back," Klaus shouted.

"I can't stand doing nothing, watching you destroying and killing yourself," Valerie was worried about Klaus.

"I love him," Klaus was in tears.

"Do you think you are doing good by doing this? He repaired your family. Please don't destroy what he built with you. He will back, I am sure," Valerie motivated Klaus.

"How do you know?" Klaus asked his mother.

"You have never loved anyone this much. Your love is pure, and God knows it. He will back here in this house within no time," Valerie said.

"Your kids need you," said Denis.

"And they need him," Klaus replied.

"And he needs you," Denis said.

"He needs me!" Klaus agreed.

"He needs you to be strong to pass this rough patch," So Denis kept on motivating Klaus.

Klaus went downstairs and looked at his kids. "It is time to have dinner," Klaus said.

"I cannot eat," Charley refused.

"We all will sit and have dinner as we always do when he was here. Nothing will change until he comes back. Am I clear?" Klaus tried to bring everyone around.

Everyone sat around the table and began to eat. Each of them was looking at the chair Hassan used to sit on. Nobody could fill his absence that had left an enormous space in their lives.

****

Chapter 21: Confession

"Life is not always about happy endings. Sometimes its about finding happiness in the ending you get" - Chip Rossetti

Klaus went for a short walk after dinner. It was raining outside, but he didn't bring an umbrella. He walked in the rain so that nobody could listen to him shout. His eyes closed, and his head rose to the sky, the tears had washed away with the rain, but the rain could not wash away his pain. He felt that he was inwardly burning; that fire could burn for him the whole world. He opened his eyes and saw every drop of the cloud fall on the ground. Then, he saw Hassan walking to him, with a smile on his face. Klaus opened up his arms broadly for Hassan. But then, Hassan stopped and looked at Klaus for a second, and then he vanished, and the rain stopped.

Klaus kept looking at his hands, and he took a long sigh; he came back home shattered with Hassan's disappearance. As he approached closer to his house, he saw some people standing in front of his house holding candles.

Klaus saw Andrew among those people, holding a candle in his hands. Klaus stood nearby and watched them. The other people followed behind Andrew and prayed for Hassan. Andrew moved and placed the burning candle on the sidewalk. That was when he noticed Klaus; he walked toward him and hugged him, and in silence, Klaus cried.

"We will find him," Andrew said.

Andrew then took Klaus inside the house. Andrew saw the piano and walked toward it. He then sat down on the bench, opened the piano, and started playing. He looked at the picture of Klaus with the children and Hassan. His closed his eyes and kept on playing. He was playing angrily and sad at the same time. Andrew's eyes were closed, and his tears hadn't stop for a second.

Everybody was angry with pain; such was the story of Hassan and his big family. Andrew kept playing, while everybody cried. Klaus looked at Andrew and felt assured that he wasn't alone in it. Everybody had his back.

Sixteen days had gone by, and Hassan was searched everywhere by the Police and his friends. Many people helped to find Hassan, but no one had succeeded yet. Klaus felt depressed and broken. Andrew hadn't been eating and felt exhausted. Ethan, Pete, Cara, Kyle, Tristan, John, Carlos, Anton, and Nina kept looking for Hassan every day but always met a dead end.

****

It was nearly 7 p.m., and it was time for Hassan's medicines. Brian gave Hassan his meds. He looked at Hassan, and saw how broken and damaged he was. A fierce darkness came over him.

***

Andrew opened his eyes and saw Hassan sitting beside him, playing and pressing the keys to tease Andrew. Andrew rubbed his eyes to see if it was real and that Hassan was back. As he approached him, he disappeared into thin air.

***

In the dark dungeon, the chains started moving, and Hassan was lifted. Hasan fell into the water as the chains stopped. He wasn't resisting this time or even moving. In that state, he saw his dad cry for him; he could not breathe seeing what was in front of him. His eyes were shut, and he was ready to die this time.

"Sorry, Klaus, I cannot be with you," Hassan silently apologized to Klaus.

"It is okay. This is the end," Hassan told himself.

Hassan felt something pull him out of the water. He opened his eyes to see it was Brian but having seen his father crying had drained Hassan of his remaining energy and he began to lose his consciousness. In that moment, Hassan felt Brian holding him in his arms.

"This has to stop now. You are killing him," Brian yelled.

The man with the hat left the dungeon. Brian looked at Hassan and said, "I will save you. Tomorrow you will be free."

He then let Hassan sleep between his arms the whole night. In the morning, Brian prepared breakfast for Hassan and sat down to talk to him.

"Today, you will be free from all this madness. You will go home," Brian assured Hassan.

Hassan cried and tried to speak, but he couldn't.

"Today, I won't stable the chains in the hooks. Always remember that I love you. Goodbye." Brian said.

He then kissed Hassan and left, who had so many questions in his mind to ask Brian, but Hassan knew he wouldn't see him again. Hassan was crying and continued to cry until he lost consciousness.

Brian arrived at work and went directly to Ethan's office. "I need you to call Andrew right now," Brian commanded Ethan.

"Hi, Brian. What do you need from Andrew?" Ethan said to Brian.

"Call him now," Brian said to Ethan again.

"Not before telling me why you want me to call him?" Ethan was adamant about getting to know the whole situation.

"It's about Hassan. Call him now," said Brian.

Ethan called Andrew, but he did not reply. So, Ethan left a message for him, and as soon as he read it, he called back and spoke with Brian, who told him everything. After the call, Andrew quickly changed to put on two different shoe pairs. He didn't notice that because his first priority was finding Hassan.

At last, they had found him, and knowing that he was alive had brought a wave of relief for them. Andrew called the Police and notified them of Hassan's location, asking them to bring a paramedic with them. He opened the door outside and saw Julian pressing the bell.

"What are you doing here?" Andrew asked Julian.

"I thought I can come by and check on you. Any news about Hassan?" Julian asked.

"We found him with the help of Brian. He is Cara's brother. I have to go," Andrew told him.

Andrew hugged and bade farewell to Julian. He went all the way to an abandoned warehouse with the Police and the paramedic. They broke in, and Hassan was found. Andrew ran up to Hassan and hugged him, glad to have found his best friend. But then something unexpected happened, Hassan pushed him away, yelling, "Please stop, I can't take it anymore." Andrew was shocked; he couldn't understand what was happening.

"Hassan, it's me, Andrew," Andrew tried to wake Hassan up.

Hassan finally opened his eyes and saw Andrew; he couldn't hold himself back, so he threw himself on Andrew's chest, and screamed, "Ahhh!"

"It's okay. You are safe now," Andrew said hugging him back.

****

Hassan was released from the chains by Police and received treatment by the paramedics. Hassan looked at his hands, took a long sigh, and lost consciousness after being released from the chain. Hassan was brought to the ambulance by the paramedic. Andrew called at Klaus' house, and Edouard replied. Andrew told him Hassan was found, and he would text the details to Klaus... Edouard ran to his father's room to tell about Andrew's call.

"Please tell me you are not joking," Klaus was astonished.

"They found him. Check your phone," Edouard said.

Klaus checked his phone and read Andrew's messages. Then, he ran as fast as he could to his car. Charley saw Klaus running and asked, "What is going on?"

"They found him." Edouard told Charley.

"They found him?!" Charley screamed a bit.

Charley shut her eyes and sat down on the stairs, taking long sighs. "They found our dad," Edouard jumped with happiness. María and Adrien listened to his happy squeals and they all thanked the Lord.

Ethan had texted Pete, Cara, Kyle, John, and Carlos that Hassan has been found and the gang finally felt they could breathe as relief took over them.

Cara received a call from an unknown number. She responded, and she was told that her brother, Brian, died in a car crash.

Cara ended the call, and Kyle asked her, "Are you okay?"

"My brother died in a car accident," Cara could not hold her tears, and she went in a state of shock.

Kyle hugged Cara and said, "It's okay. I am here with you."

She found a voice message from Brian. Cara looked at her phone. She did not open it. She was scared of what she would hear.

****

When Hassan woke up, he began to yell, "Leave me alone," to a nurse who was treating Hassan's injuries in the hospital. Hassan had been aggressive to the nurse. Klaus was standing close, observing him. He saw that Hassan had been severely injured. The doctor asked for Hassan to be held in the hospital bed to ensure his safety.

"No," Klaus denied the doctor.

"Do you want him alive and us to treat his injuries?" the doctor asked.

"I do," Klaus replied in the affirmative.

"We have to restrain him to his bed," The doctor told Klaus.

"Do whatever it takes to treat him," said Klaus.

The doctor and the nurses restrained Hassan to his bed. Hassan begged. "No. please don't do this to me."

Klaus cried seeing Hassan like that. Then, Hassan looked at Klaus and said, "Please don't do this to me."

Andrew hugged Hassan and said, "It's all right. It's provisional." Klaus leaned on the wall and then fell to the ground. In that moment, Klaus realized he was back to square one; he was yet again unable to hold back his tears. His family had been broken again, and nobody could fix it this time. The only one who was able to do it was Hassan, who himself was now too broken to mend anything. After a while, Cara, Pete, Kyle, John, and Tristan came to see Hassan. They came into Hassan's room and began talking, trying to cheer him up and distract him from the trauma. Hassan listened to them talk grinding his teeth. He moved his hands and body in a way that made him scream in pain. Then he shouted, "Shut up. You are all so loud!" And broke into tears.

Everyone was looking at Hassan with tears in their eyes. Ethan moved close to Hassan and said, "What have they done to you?" He put his hands on Hassan's shoulder, but Hassan tried to avoid Ethan's touch. Everyone was shocked to see Hassan's behavior. Andrew ran out of the room, crying, and John followed him.

"Are you alright?" John asked Andrew.

"I am not alright. He is not alright. We all are such a mess. I can't see him like this. I can't… I can't breathe," Andrew said while crying.

"I believe everything will be back to the way it was before this."

"I don't think so. Brian said that Hassan has been tortured in horrible ways every day."

"Brian!"

"I am afraid he is gone."

The nurse entered the room. "Please keep in your mind only two people are allowed," the nurse said.

"Any updates?" Andrew asked the nurse about Hassan's condition.

"We have to take him to radiology to check his injuries. It seems his right shoulder is seriously injured. He might have to undergo a surgery. Also, we need to check his viral loads," The nurse replied.

"Can you give him anything to sedate him?"

"Let me ask the doctor."

Andrew walked to Hassan's bed. "You will be okay," he said.

"I want to close my eyes and just be gone," Hassan cried.

"No, you don't. Look at him, look at Nicholaus. Do you think that if you were gone, we would be okay? Nicholaus will kill himself, and I cannot live without you," Andrew replied to Hassan.

"Don't you see, I am defeated and damaged; I feel like a shadow. You all need to let me go," Hassan could not get himself together as he was scared of the torture.

Klaus got up and walked towards Hassan's bed, crying. "You have no right to push us. You have no right to give up. And if you do, you will not only destroy us but kill me as well. I don't want to live in this world without you. You need to fight, and we will all fight with you," Klaus said.

"I have lost my fight," Hassan was discouraged.

"Then stand up and fight again," Klaus motivated Hassan.

"Fight for us," Andrew followed Klaus' lead.

Because of the tears in his eyes, Hassan could not see well. Hassan moved his eyes and watched Brian hold his hands. He looked back.

Brian smiled, "They are right. You need to fight," he said. Hassan took a long sigh and asked, "Is Brian dead?"

"He sacrificed himself to save you," Andrew told Hassan."

"Where ever I go, death follows me," said Hassan.

"That is not true" Klaus disagreed with Hassan.

"Don't be damned. Nicholaus is here. Let him hold you between his arms to feel that you are alive," Brian talked to Hassan.

Klaus held Hassan and kissed him on the forehead, "I love you," he said.

****

Hassan felt a strange feeling that something in him allowed fear to control him. Hassan looked at the bed around his feet, and he saw himself sitting on the edge of the bed with an evil look.

"Do you remember me? I am your fear. You never accepted me as part of you; you abandoned me. Now, you pay the price by being miserable and lonely. I will make sure to destroy everything you touch or love. You can't run, you can't hide or even fight. You are just a shadow, an empty shell. Such a waste you didn't accept me along with your dark side. You had to be that angelic ass so that everyone would love you. That desperate need for love is so pathetic, but it is too late now. You are too consumed by me." Hassan closed his eyes as he was too distraught to confront his shadow.

"Before I go, I need to ask you something. Why were you afraid of me when I found you?" Andrew asked Hassan.

Andrew was confused by Hassan's reply, and he thought, "How is that possible? But maybe, it does make sense."

Cara, Pete, Ethan, John, Kyle, and Tristan had gone home, and Carlos was already there. Everybody noticed John was furious.

"What is your problem?" Ethan confronted John.

"Cara, is there anything you want to share with us?" John diverted everyone's focus to Cara.

"I don't understand. What do you mean?" Cara asked John.

"Did you or did you not know that Brian is the one who kidnapped Hassan?" John kept on charging on Cara.

"I have no idea what you are talking about. Anyway, I haven't spoken to Brian since Ethan's birthday. Even the voice message that he left me today, I didn't open it. And he is dead," Cara tried to justify.

"Can we all hear that voice message?" John asked.

"Her brother just died, and she didn't do anything wrong. Be respectful," Kyle tried to stop John.

"I want to hear it," However, John was stubborn in probing into Brian being behind Hassan's kidnapping.

"Fine, I will play it." Finally, Cara gave in to John.

Cara played Brian's voice message. Brian sounded like he was crying during the voice message. "Hi, Cara. It's me, Brian. I am sorry that I hurt you in many different ways. I am sorry that I have caused you unmeasurable pain. I regret everything I have done to you and Hassan. I took advantage of him and destroyed his life. I love you. I know I wasn't the brother you always wanted, but you have Hassan now. He is exactly the kind of brother you always wanted and so rightfully deserved. I saved him today, and I know that will cost me my life, but I don't care. I love him, and it is time to set everything back to the way it was. Please take care of yourself. Goodbye." This was the message Brian had left to Cara. After they all heard the message, John stormed out.

"How could you do this to us. You brought him to our life. I was happy with him. I had a perfect life with him. I will never forgive you for what you have done to us. I will never forgive you," John yelled at Cara.

"It always will come to this? All our time together was a waste because you are still in love with him. We are done!" Carlos cried. Carlos walked away and left the house.

John looked at Cara, "I hope you are happy now," John said and went to his room.

"Don't listen to him. He is angry," Ethan said.

"He is right. It all my fault. I brought him here, and I knew what he was capable of," said Cara.

"Don't blame yourself," Kyle tried to console Cara.

She looked at Kyle and said, "I am the one who destroyed him. It's all my fault."

"You didn't know what would happen. It's not your fault," Kyle tried to reason with Cara.

"I knew better. You are better off without me," Cara was disheartened and took the blame for everything.

"What? How could you say that?" Kyle asked.

"I can't be with you," Cara intended on ending things with Kyle.

"Are you insane?" Pete could not believe what Cara had just said.

"What is wrong with you?" Tristan felt the same as Pete.

"It's not your fault. Whatever Brian did is on Brian's. Don't blame yourself. You didn't know. Hassan broke up with John because John didn't trust him. We all know that," Ethan tried to bring Cara around.

"It all started when I invited Brian to this house. So, stop defending me," Cara vehemently put a deaf ear to everyone and blamed herself for how the things had transpired.

It was an indescribable moment for Cara. The pain she felt at this moment was different, more unbearable, more soul-crushing. The vicious, stinging flavors of hopes were torn up, and she couldn't shrug it off because she had lost everything today.

****

Kyle left the house. His heart felt halted. He shouted and wanted Hassan to help him out. He couldn't believe what Cara had said to him. He was not only hurt, but he also felt like he was being buried alive. He loved her profoundly. He walked and cried in the street, but Hassan wasn't there for him this time to help him heal.

Andrew returned home while Hassan was still in the hospital with Klaus. He found Ethan holding Pete and Tristan when he entered the room.

"What is going on?" Andrew asked.

"John had a fight with Cara. Carlos left John and Cara broke up with Kyle," Pete described the situation to Andrew.

"We need him back," Ethan referred to Hassan's absence.

"That won't happen soon," Andrew told Ethan.

"What do you mean?" Tristan asked Andrew.

"He will have surgery today. After that, they might admit him to psych." Andrew told everyone about Hassan's current health condition.

"They will admit him to psych?!" Ethan did not take the news lightly.

"You all saw him today. He is broken, damaged and defeated," said Andrew.

"Will he ever be the same again?" Pete asked.

"I don't think so," Andrew answered Pete.

"Why did Brian do this?" Ethan still could not understand Hassan's kidnapping.

"Brian saved him. The mad man did this," Andrew tried to clear the confusion.

"The mad man!" Andrew was puzzled.

"Only Hassan can expose him," Andrew said.

Andrew was thinking about what Hassan had told him at the hospital. He kept repeating everything in his mind to remember the details. Then, finally, he took a long sigh, "I don't know what I should do," he said to himself. Andrew went to his room, took Mike's painting, and talked to it. "I gave up counting on days because I am fed up. I am tired, and I can't help him. To be honest, I don't know what to do. It is time for you to help him, guide him."

****

Hassan's surgery was about to take place. With him were Klaus and Andrew. They had anesthetized Hassan, and the countdown started. The surgical team was operating on Hassan. Hassan was watching the surgeon cutting into his flesh at the surgical table when somebody called him; he turned to see Mike standing in a corner. Hassan hugged and ran into him.

"I've missed you," Hassan said out of joy.

"I've missed you too," Mike shared his mutual feelings with Hassan.

"Wait, Am I dead?"

"No, you are alive."

"Why are you here with me?"

"I cannot move on before I am sure that you will be okay."

"I don't think I will be the same. I am broken."

"What happened to you?"

"I lost a big fight."

"Your fight isn't over."

"What do you mean?"

"He will come back to you and for your beloved ones. Are you okay if he hurt them the way he did to you? Will you watch him and do nothing? You always wanted to protect them, but now you need to decide before it's too late. And if they lost you, they won't recover from it. It's now up to you. Do you want to finish your fight, or will you be that coward who couldn't fight?"

"I want to, but…."

"Isn't that enough, or do you want more?"

"I don't know if I can fight anymore."

"Since when are you a coward?"

"All I want is to die and reunite with my family."

"What about them? Aren't they your family too? What about your kids? What will Klaus tell them?"

"They are his kids."

"They are your kids too, and you have two more coming on the way."

"Two more!"

"Yes, a boy and a girl. They will be exactly like you. What a shame you won't be around since you've given up everything, and for what?"

"I cannot fight anymore."

"You can, but you don't want to."

"I can't beat him," Hassan pointed to his reflection.

"You already know what to do to beat him. The best way to beat your fear is to be it. You need to accept that part of you," Mike tried to comfort Hassan.

"I can't."

"Why not?"

"I will hurt them, and I don't that to happen."

"You need to decide who you are, and when you do, you will know how to protect them. Believe me when I say this, he doesn't have a chance to beat you but that'll only happen if you accept that part of yourself," Brian spoke up.

"He is right; he doesn't stand a chance against you," Mike seconded Brian.

"You are a warrior. Just follow your instincts," Brian said.

"Do you love Klaus, your kids and your family?" Mike asked Hassan.

"I do," Hassan nodded.

"Then don't be that coward. Rise and shine, son. This is your fight, so get your mind and armor ready because things will be rough," Mike asked Hassan to gut up and face his fears.

"Remember, you are not alone in this. You have them, and they are alive. Also, you have us to support you, not to guide you. It is your fight, and you have to decide for yourself," Brian showed his support to Hassan.

"We are all here for you," Hassan's father joined in too.

"I chose you because you are the only one who can repair my family. We all know you have it in you," Lizzie came in too.

"It would be best if you unleashed it; embrace it," Hassan's mother said.

"We don't give up," said Hassan's sister.

"Don't give up, brother," Hassan's brother stood alongside him.

"I am here because of you. It is time to let me go," Nie appeared in front of Hassan.

"We will leave you to decide. Take your time, but remember the darkness is coming for you," Brian left the decision to Hassan.

All of them left. Hassan looked at himself at the operating table, and the surgeon had completed the surgery by then. In the hospital corridors, Hassan moved from room to room, looking for an answer everywhere. Instead, he just saw sorrow and pain. When he saw a man standing and watched himself dead, he stopped near a room.

Hassan asked him, "Are you alright?"

"I am dead. How Am I supposed to be alright?" the dead man replied.

"I am sorry," Hassan apologized to the dead man.

"I will do anything to feel that pain again. You still have your chance," said the person Hassan was talking to.

"I don't know what to do?" Hassan was clueless.

"Life is pain, and you are used to that pain. If you die, you will kill him," the person pointed to Klaus. "You need to go back, embrace what you have and fight. You don't belong here." The person told Hassan. Then, Hassan ran in and woke up in his room.

Time went by so quickly, but there were still a lot of things that Hassan didn't talk about. Everybody at home was safe. Since that second day, Hassan hadn't seen Cara, Kyle, John, Carlos, or Anton; he had been in hospital for two weeks. Every time he asked about them, he didn't get a reply. His family had been completely broken, torn up, and nobody could fix it. It was finally time to go back to the house. Klaus and Andrew took Hassan home. Hassan sat by the window and looked out. Tristan, Valérie, Denis, Nina, Evelyn, Ethan and Pete were there, all of them were waiting for him. After Klaus, Hassan entered the house.

Everybody was glad he was back, but was he? Hassan looked around and saw Lizzie, his father, mother, brother, Nie, Mike, and Brian standing in front of him. They all went to him together and began to talk. Hassan was breathing heavily, panting, grinding his teeth, and said, "Too loud. Too loud," and walked away quickly. Klaus said, "Wait," and held his right hand. Hassan screamed in pain and got down on his knees, and crawled to hide behind the couch. Everyone was painfully watching him.

"No, not again," Andrew said.

"Nobody talks to him or get any close to him," Klaus directed everybody in the house.

"Did Hassan give up on himself? What is happing to him? Will he be ever the same?" These questions were in everyone's mind. María forgot she had left Adam on the floor. Lizzie looked at Adam and said, "You are my sister's son, and I need your help."

Lizzie took Adam's hand and walked towards Hassan. Until he touched Hassan and said "Dada," no one had noticed Adam. Everyone was all pale, frightened by what was happening.

"Come here, Adam," Klaus said.

Everyone in the house stopped at that moment. Hassan's eyes were closed, "I'm a dad. The first word of my son is Dada, but I'm doing what? My struggle is not over, until I say so. The time has come to make a decision. Am I cowering that? Or, perhaps I can't control myself. I will accept that part of me," Hassan said.

Andrew tried to take Adam away quickly. They all were scared of the aggressiveness of Hassan. Hassan turned to Adam and hugged him. As he was frightened, Klaus closed his eyes.

Hassan smiled with tears in his eyes and said, "Yes, I am dada. I missed you so much. Forgive me, dada is a little sick."

Klaus opened his eyes and saw Hassan kissing Adam on his hands and forehead. Everyone was crying at the sight of old Hassan. Hassan stood up and held Adam between his hands and took a long sigh.

"You are back," Andrew smiled with tears in his eyes.

"I forgot who I was in the middle of a fight. I thought the fight was over, but I was wrong. The fight is still on, and I have to win because I am a warrior." Hassan's hands were shaking.

Hassan wiped his tears and said, "I am here. I can't hug you all because of my surgery. Let's sit and chat," Klaus looked at Hassan and said, "You are here!"

"Yes, I am, and you need to kiss me to remind me how it feels to be alive," Hassan said to Klaus.

"I've missed you so much," Klaus said. Hassan kissed Klaus and said, "I am here now." Hassan looked at everyone.

"Let's have a seat and talk. María, have a seat with us," Hassan said to everyone.

"Not before I make juice for all of us," said María.

"Thank you," Hassan said to María.

"I missed you, dad." Adrien came in to see Hassan sitting on the couch. He ran to hug him.

"I am happy you are here sitting between us," Edouard said to Hassan.

"Me too," Hassan replied to Edouard. "Charley, where is Anton?" Hassan asked Charley.

"We broke up," Charley told Hassan.

"Let me guess, he said something about Peyton, and you told him that he is not her father," Hassan inquired Charley.

"Yes, I told him he is not her father," Charley nodded.

"Well, you are wrong because he is her father. Call him now and ask him to come," Hassan was back in his dominant character in no time.

"But…" Charley was about to argue, but Hassan interrupted her.

"You will never find a good father like Anton. Call him." Hassan was not taking no for an answer.

"I will now," Charley agreed to Hassan.

"Adrien and Edouard, we will talk every day like we used to. Now Andrew, what are you hiding from me?" Hassan asked Andrew.

"Cara broke up with Kyle, and Carlos left John," Andrew gave the main news to Hassan.

Hassan took a long sigh, "Call Cara and ask her to come here. Don't say anything else to her. I will call Carlos and John. Tristan, why didn't you go out with David?" He inquired after giving instructions to Andrew.

"We were all worried about you," Tristan replied.

"I was in the hospital, and that shouldn't stop you from getting your happiness. Pete and Ethan, I assume you also had a fight." Hassan was gradually getting back to normal and having conversations with everyone, but at the same time, someone pressed the doorbell, and everyone's eyes went toward the door.

****

Chapter 22: Mad-Man

"Never underestimate the power you have to take your life in a new direction" - Germany Kent

As everybody heard the doorbell ring, Klaus decided to attend to whoever was on their doorstep.

"I will open the door," Klaus said.

Hassan continued to talk to Pete and Ethan. Then, Klaus opened the door, and it was Cara.

"Cara, we were just talking about you. Come on in," Klaus welcomed Cara into the house.

"Klaus, I will be quick. I am here to drop off my keys, and I want you to tell Hassan that I am leaving," Cara said to Klaus.

"No, please. I just got him back. I can't tell him that," Klaus refused to do what Cara was asking of him.

"Tell me what?" Hassan overheard Cara and Klaus and then joined in.

"I am here to drop off your house and restaurant keys. I am leaving today," Cara broke the news to Hassan.

"Come in," Hassan brought Cara in.

"I have to go. I don't want to miss my train," Cara was in a hurry.

"I said, come in," Hassan yelled.

Cara entered the house with her suitcase. "You are leaving?" Hassan asked.

"Yes, I am," Cara nodded.

"Does Kyle know you are leaving?"

"We broke up."

"Klaus, can you give me the phone, please?"

Klaus gave Hassan the home phone, and Hassan dialed Kyle's phone number.

"I have to go," Cara insisted.

"I am not done yet," Hassan said.

"Hello!" Kyle said, answering Hassan's phone call.

"Hi, Kyle. It's me, Hassan. How are you?" Hassan greeted Kyle.

"I am glad to hear your voice. I am okay, and how are you doing," Kyle asked how Hassan was doing.

"I am doing fine. Do you know Cara is leaving today?" Hassan asked Kyle if he knew about Cara's departure.

"No, she won't talk to me," Kyle replied in a disappointed tone.

"Well, you need to be here with her," So Hassan called Kyle over to his place.

"She doesn't want to be with me," Kyle was adamant.

"Do you want your child to grow up without a father?" Hassan asked Kyle.

"How…" Cara was interrupted.

"Is she pregnant?" Kyle wasn't aware of Cara's pregnancy.

"With a boy. You need to come right now," Hassan confirmed.

"On my way," Kyle instantly hung up.

"See you then," Hassan responded.

Hassan ended the call and looked at Cara.

"I caused all of this to you, to John," Cara said to Hassan.

"You didn't cause anything. I hear John talking, am I right?" Hassan tried to convince Cara.

Hassan walked to Cara and said, "You have no right to think like that or even think of leaving. And you absolutely have no right to take your son's family away from him. He will grow up between us," said Hassan.

"But..." Cara was interrupted by Hassan.

"That's not your decision alone. It must be awful to have all those painful thoughts. You are pregnant; you need to rest," Hassan said to Cara.

"How did you know I am with a child?"

"Your smell is different, and you are glowing. Come, have something to eat. Brian must make you hungry all the time."

"What did you call him?"

"His name will be Brian."

Cara hugged Hassan and cried, saying, "Thank you."

"Go and have a seat. I need to talk to Klaus."

"I am sorry. I didn't mean to step over. Please don't be mad," Klaus apologized to Hassan.

"Of course, I am mad, but not for what you said. I am mad because you keep apologizing. Stop apologizing. I got it, you are protective, and that is your right. I am the one who should apologize for what I put you and my family through. I stopped fighting, but now I am back, and I will fight till the end," Hassan bluntly said to Klaus.

"Will you spend the rest of your life with me as my husband?" Klaus proposed to Hassan.

"Yes! Yes! Yes!" Hassan instantly agreed.

"Three, yeses! You didn't even think."

"Why do I have to think?"

"I will be your husband under one condition."

"I agree."

"Don't you need to know what it is first?"

"I won't change my mind."

"I want you to have my last name."

"I will be honored," Hassan smiled.

"I love you," Klaus confessed.

Hassan and Klaus shared a long, deep and passionate kiss.

"I have to make some calls," Hassan excused himself.

****

Hassan then called Carlos, John, and Oliver. Andrew had also arrived at Klaus's place and said, "It is necessary for us to speak." Andrew spoke to Hassan about his comments in the hospital, and Hassan told Andrew to tell no one what he found. "You have to do exactly what I'm going to say to you. The time has come to end this," said Hassan. Oliver and Carlos were there, but John sat far from Carlos.

"For God's sake, stop being idiots. John, you need to apologize to Cara and Carlos, and Carlos, no need to run away every time something comes up. It is simple, are you in or out of this relationship?" Hassan asked Carlos.

"He's still in love with you," Carlos replied.

"Stop this nonsense. He is angry because he thought that Brian stole his life, but he forgot that he is responsible for his actions. John, you made your choice and believed I cheated on you without talking to me. You thought your life was ruined because you got HIV, but look at you, you are healthy. The only thing that will ruin your life is letting your love slip away from your fingers. Don't be damned; you have a chance to be happy with him. Carlos, you need to fight for what is your right. Now, I want you to sit with Cara," Hassan schooled everyone there.

Carlos took John's hand and sat down next to Cara. John apologized to Cara and kissed her on the forehead, and then kissed Carlos as well. After that, Kyle showed up holding flowers in his hands.

"Hi, everyone. Cara, I need to ask you a question," Kyle directed towards Cara.

"Go ahead," Cara approved.

Kyle got on one knee and asked her, "Will you marry me?" Cara, with tears, said, "I do."

"What are you waiting for? Kiss her already," Hassan told Kyle.

Kyle kissed Cara and smiled. "We will have a double wedding," Hassan said with a huge smile.

Charley, Cara, Pete, Ethan, Kyle, Tristan, John, Andrew, and Carlos together said, "What!"

"You all heard me," said Hassan.

"There will be three weddings" Anton also came in.

"Three!" Charley acted surprised.

"We are getting married," Anton said.

"But you didn't ask me," Charley said.

"He asked us, your dads and we said yes," Klaus told his daughter of his and Hassan's approval of Anton for her.

"You need to be with a real man, plus you are in love with him. Your daughter needs to be with her father permanently," Hassan said to Charley.

Hassan went to the kitchen. He walked to María, hugged her, and kissed her on the forehead.

"Thank you" Hassan appreciated María's service.

"I was worried about you," María said to Hassan.

"I know, but I am here. Klaus, we need to talk," said Hassan.

Hassan took Klaus to their room. "I need to ask you to do something for me tonight," Hassan said.

"Anything for you," Klaus agreed.

"Tonight, I want you to make love to me as you used to. Don't stop, no matter what," Hassan demanded.

"Why?" Klaus was somewhat confused.

"Do you want me back to normal?" asked Hassan.

"More than anything," Klaus undoubtedly agreed.

"Then do what I asked you."

"Can I ask why?"

"The fight is not over yet, and I need to be whole again to finish it. You will help me to be whole again. I need to face my fears. Are you in?"

"I am, all the way."

"Keep in mind if you ever stop, I might be gone forever."

"I will do whatever it takes."

"I am counting on you."

****

Everybody was happy after dinner. Hassan had gone for a shower. He stood under the water; his hands moved through the body. He took a deep breath and opened his eyes. He saw himself standing inside the shower (as his fears). "You can't defeat me. Tonight will be your end. I will finish you off." Hassan ignored his fears, closed his eyes, and heard the drops of water fall on his body. Then he shifted his hands to his chest and on his shoulders. His fears continued to speak, but Hassan kept on ignoring them. Finally, Hassan finished his shower, wandered naked to his bed, and dried himself. Klaus was smiling, taking a deep breath and saying, "I have lost so much to you," Klaus looked at Hassan and said, "Are you ready?"

"I am," Hassan gave a slight smirk.

Hassan laid down on the bed, and Klaus started to kiss him. "Don't ever stop, no matter what," Hassan said.

"You have my word," Klaus said.

Klaus kissed Hassan all over his body. Hassan looked up and saw his fear on top of him, locking its gaze on him. "Ready!" Hassan nodded his head. The fears whispered in Hassan's ear, "Now you are chained, standing on the cliff about to fall." Hassan closed his eyes and stood chained on the edge of the cliff, and there were fears. He was driven to fall by his fears from the cliff. Before falling, Hassan held his fears by his hands.

"What are you doing?" fear asked.

Hassan answered, "You are a part of me; if I am going down, I will take you with me.".

"You are going to die." Hassan fell, and his body trembled. So, his tears continued to fall from his eyes. Klaus knew he couldn't stop. Hassan bumped into the water, breathing heavily into his body. With every second, he drowned deeper as the chain pulled him in deeper. He drowned continuously until he reached the ground of the water and then saw the large sword slipping his flesh.

"Why can't I move?" Hassan said.

His fears replied, "I told you; you can't defeat me."

Hassan couldn't move, "I need to move, damn it," Hassan said.

Klaus started being worried. Hassan hadn't shown a movement.

"Please come back to me.", Klaus said, but Hassan didn't move. Klaus shouted, and his tears fell on Hassan's face.

Something happened that Hassan couldn't understand. Hassan felt the waters had changed. "This is your end," he heard a voice saying.

Hassan was unable to understand his sensations and then looked into his fears eyes deeply. This was a sad and silent moment. Unable to conclude his battle, Hassan wouldn't give up on him but Klaus. Klaus continued to whisper and cry.

Hassan smiled and said, "Now I understand." Hassan pulled his fears into him and absorbed him. He closed his eyes and said, "I will never abandon any part of me again. Not my fears, not even my dark side."

The water led Hassan to the surface, looking up and taking a deep breath. He looked at his chains before reaching the surface, and the chains were missing. He picked up the sword and drowned it in the water. Finally, he reached the surface and could breathe at last. Hassan breathed deeply, and Klaus flipped up onto his bed. Hassan said: "I love you." Klaus then said, "Please do not leave me."

"Do you think I will ever leave you? You are the love of my life. I would destroy the whole world for you," Hassan said.

"I love you," Klaus said in tears.

Hassan passionately kissed Klaus. The kiss was blended with Klaus' tears. It tasted like tears and love. That taste of love that never would die and every day would become sweeter and more robust. They kissed for a while, then Klaus stopped and said, "Something's different about you?"

Hassan looked at Klaus with a smile and said, "I am the same person you love. And I am deeply in love with you," They continued to touch and kiss each other. They melted in each other and fell in love all over again.

"I've never done it before with anyone, and I want you to be my first," Klaus told Hassan.

"You don't need to do that. I am okay with what we have," Hassan replied.

"You are the one, and I want to have it."

Hassan bit his lip, "Are you ready for the ride?" Hassan said.

"I am ready."

"I will take you to the clouds for an epic ride. I am all yours, and you are mine. I will never stop loving you."

Hassan bit Klaus' ear lobe and put his tongue in his ears. Klaus felt something different about this Hassan, He was no longer innocent, but he remained the same sweet person. This time Hassan had taken Klaus on another ride, but it was all different. More than ever, Hassan made him lose control and added a sweet pain to pleasure. He closed his eyes then opened them to feel he was on Hassan's clouds.

Hassan smiled and asked Klaus, "You're about to ride again?" Klaus watched and smiled at Hassan. Klaus was taking Hassan between his arms and flew among the clouds. He was opening his black wings.

"You don't want to miss the view from here," Hassan smiled and spoke. Klaus smiled and said, "I love you" Hassan wrapped up his wings around Klaus and himself, sopping with a pot. Hassan whispered in Klaus' ear and he smiled with closed eyes.

Klaus breathed deeply and liked what Hassan was doing. Hassan gave a kiss to Klaus, then whispered again in Klaus's ear. After some time, Klaus opened his eyes and wrapped his arm around Hassan's waist, and the other arm moved around Hassan's back. Then he said, "I love you." Klaus looked deep into Hassan's eyes and he replied with a sweet voice. "I love you too."

****

The following day, Hassan went downstairs and sat down on the chair and looked over to Klaus' office window. For a while, he didn't move. Then, he began to remember and smiled on the part of his life that he had spent with John. He continued to look at the window and decided to remember each detail of his life with Klaus. For about thirty minutes, he didn't move from the chair. Then, the door suddenly opened, and he heard Andrew say he would talk to Julian. "Please help him," Andrew said with pain. Julian said, "I will try. Now leave us alone." Then Julian closed the door. Julian moved closer to Hassan.

"I can recognize that perfume. It is a symbol of my pain and torture. Welcome, Manauris," Hassan stood and turned to confront Manauris.

Manauris took off his glasses and smiled.

"I would be really disappointed if you didn't recognize me. Finally, you opened all doors of your mind together to walk into all places in your mind to defeat your own fears. You are unstoppable now. I waited so long to find you. You are my perfect creation," Manauris said and took a long sigh.

"Am I?"

"I knew it since I saw your paintings, especially the one that Brian has stolen. Look at you now. You don't need help."

"So much for your theory to put the person in an extreme amount of his fears and force him to face his all fears to be free. To walk into people's minds and open every door and find what's behind them. Don't you think it is too much?"

"It doesn't work for everyone, but in your case, you are very perfect for my theory. I thought you gave up, but I can see you whole again."

"And you are here to finish your job, or there is more?"

"We are done, but I cannot leave you here. You belong with me."

"And if I said no?"

"I will destroy everything you touch or love."

"I like you already. But I am afraid I am a little bendy, and you cannot fulfill my desire. You awake a dark angel who is hungry all the time. Thirsty for more and more. Maybe you can do something for me."

"What do you mean?"

"So, tell me, how did you kill Brian? No, wait, let me guess. You killed him using his car, so it looks like an accident. Am I right?"

"He got soft because of his love for you. He ruined everything we had worked for."

"So, you killed him because you think that his love and feelings are worthless?"

"The whole point is being free from your silly emotion. You were his weakness, and I tried to destroy you. But here we are."

"I am impressed. So, anyone messes with your process, you would destroy him. How many times have you done this?"

"I am a scientist; I make sacrifices."

"You killed Brian, ruined many people's lives, including mine, put me through hell, yet it's not enough. You disposed us like garbage, like we were nothing. You are not a scientist. You are just an asshole who abuses and messes with people's minds. You are nothing, nobody," Hassan screamed with tears in his eyes.

"I guess I have to finish you," Manauris was determined.

Hassan laughed very loud, then looked at Manauris and said, "I am scared," with mocking facial expressions.

"You cannot defeat me."

"Sorry, what? I couldn't hear you well. You are wrong. I cannot see you because you are so little, and I feel sorry for you. But I promise you one thing, that I will undo everything you have done and destroy you, and no one will remember you. Oh, if you don't believe me? I can prove it to you."

Hassan opened the door, and Oliver walked in with a smile on his face.

"Hi! Manauris. Do you remember me? I am Max, but now I go with Oliver," Oliver introduced himself to Manauris.

"What is this?" Manauris was puzzled.

"It is basically a freakshow created by you," Hassan pinched Manauris.

"I shall destroy you and everyone you love," Manauris threatened Hassan again.

Hassan looked at Manauris with a grand evil look and told him, "If you ever come close to my family, I will unleash my hell on you. Believe me, you don't want me to because you won't last a day."

Manauris felt that Hassan wasn't alone. He felt the presence of many people standing behind him. And what about his face? Is it a human being? For the first time in his life, Manauris felt very frightened. Tasting his own toxic medicine was an indescribable moment. He began to laugh insanely.

"Now, say hello to the cops; they will take care of you. Good luck with your new miserable life, and enjoy the view," Hassan had called the cops to get Manauris arrested.

****

The cops took Manauris. Andrew entered the room, "Finally, we will be able to move on in our lives. Isn't that right, Oliver?" He said.

"Don't mess with me," Oliver replied to Andrew.

"Hassan, Are you alright?" Andrew asked Hassan.

"I am fine," Hassan replied in affirmation.

"Something is very different about you. I can't put my finger on it," Andrew tried to identify some peculiarities in Hassan's changed behavior.

"Don't spend time thinking about it, just spend time with someone you love," Hassan responded to Andrew.

Andrew moved to Hassan, took Hassan's hand, and put a big black feather in Hassan's hand.

"Your secret is safe with me," Andrew said and walked away.

Hassan smiled and said, "Clever boy."

Sitting on the desk, Hassan opened the drawer, took his little box, and put it on the desk. Hassan continued to look at the box but did not move it. Finally, he closed his eyes and he took a long sigh. Klaus went to Hassan and stood behind the chair. But he did not open his eyes. Hassan smiled.

Klaus whispered in Hassan's ear, "I had a dream about you last night."

"What was your dream?" Hassan asked.

"You and I on the clouds, and you have black wings," Klaus replied.

"Sounds quite the dream," Hassan was fascinated.

Hassan opened his eyes, rose and wrapped Klaus in his arms, and said, "You want to dance with me?"

Klaus smiled, took Hassan's hands, and began to dance. No music, just two people in love. The sound of heaven on earth were two heartbeats; in this world, two souls from two different worlds met. Two souls that were deeply in love. Hassan and Klaus danced together for a long time. Hassan had placed his head on Klaus's chest and heard his heartbeat.

Klaus took a long sigh and closed his eyes. Hassan smiled and requested Klaus to open his eyes. Klaus opened his eyes and saw Hassan dancing on the clouds with his enormous black wings. He looked at Hassan's face and smiled.

"I love your smile, and I love you," said Klaus.

Hassan wrapped his arms and wings around Klaus and replied, "I love you."

"Now we can go on in our life," Klaus said.

"Not quite yet," Hassan stopped Klaus.

"What do you mean?"

"I have to do one more thing, and I need you to be with me."

"I will do anything for you."

"Can we go to Lake Ontario Park?"

"Lake Ontario Park!"

"You will know when we get there."

"Let's go!"

****

Picking his little box, Hassan took Klaus' hand and walked together. Klaus walked up with Hassan to the lake and sat alongside him. Hassan began to speak. Initially, Klaus thought Hassan was going to talk to him, but he didn't, so he got a little bit sad. A tear dropped to his cheek from his eyes.

Klaus looked at the sky and said, "Not again. I just got him back." Klaus closed his eyes, took a long sigh, and opened his eyes. He saw Hassan open his box, took a wristwatch, and put it on the bench, then said, "It is time to let you go." Suddenly, Klaus saw a man sitting next to him. The man smiled and said, "I know you will take care of him. I know you are the one."

The man walked then started to disappear. Hassan stood up and walked to the water and dropped his box into the water. "It's time to move on." Hassan said.

Hassan turned, looked at Klaus with a smile, and said, "Can we go home?"

Klaus smiled and whispered, "I love you."

Hassan smiled and whispered back, "I love you."

****

Manufactured by Amazon.ca
Bolton, ON